Law and Practice for Architects

Law and Practice for Architects

Bob Greenstreet

Karen Greenstreet

Brian Schermer

Architectural
Press

Routledge
Taylor & Francis Group

LONDON AND NEW YORK

Architectural Press is an imprint of Routledge

2 Park Square, Milton Park, Abingdon, Oxon OX14 4RN
711 Third Avenue, New York, NY 10017, USA

Routledge is an imprint of the Taylor & Francis Group, an informa business

First published 2005

British Library Cataloguing in Publication Data
A catalogue record for this book is available from the British Library

Library of Congress Cataloguing in Publication Data
A catalog record for this book is available from the Library of Congress

ISBN 0 7506 5729 4

For information on all Architectural Press publications
visit our website at www.routledge.com

Contents

List of AIA documents

All forms reproduced by kind permission of The American Institute of Architects, www.aia.org.

Preface

Many architects cringe when discussing issues related to the law and practice procedures because they associate these with an almost Pavlovian response to disputes, wrangling with lawyers, and threats to their livelihood. The authors of this book, however, feel that such a reaction is largely unwarranted. Far from being a source of threat and fear, knowledge of law and practice provides a welcome measure of security and certainty.

In everyday practice, the architect spends considerable time carrying out various administrative tasks and dealing with problems and situations arising from the design and construction of each new building project. In order to do this effectively, a basic knowledge of all the relevant procedures involved is necessary, coupled with an understanding of the broader legal and professional issues at stake.

Law and Practice for Architects provides a comprehensive, concise, and simplified source of practical information, giving the reader a basic legal overview of the wider principles affecting the profession, and concentrating on the more specific procedural aspects of the architect's duties. In addition, it contains a series of checklists, diagrams, and standard forms which provide a quick and easy reference source.

Each section of the book culminates with a short commentary on the architect's responsibilities entitled 'Practice Overview,' based on a series of articles published in the architectural journal *Progressive Architecture* by Bob Greenstreet. Each is followed by a Question and Answer page, addressing common problems or issues likely to be encountered at each stage of the design and construction process. Neither the Practice Overview nor the Q & A sections are intended to provide a specific answer to a problem, as each practice situation would, in reality, merit its own unique handling. Rather, they are meant to convey an attitude appropriate to successful practice management.

The most recent AIA standard forms for design, construction and construction management have been referred to extensively throughout the text. Many of the forms reproduced in the book are published by the American Institute of Architects. While their use is by no means mandatory, they are useful in providing a consistency of understanding on each project between the various parties, and are therefore recommended where appropriate.

Law and Practice for Architects offers only an introductory framework of information, as a detailed analysis of all relevant aspects of the subject could not possibly be crammed into so few pages. Many elements of law vary from state to state and, in some cases, from city to city, so it is important that readers use the text as a basic overview of the subject, checking for more detailed information where appropriate. For example, for out-of-state practice it may be prudent to investigate such information as licensing, codes, lien law, partnership laws, etc., before providing professional services. Similarly, it is not the intention of the authors to provide a legal service in the publication of this book, but to offer an introduction to legal and practical matters concerning architecture. Legal assistance is strongly advised where appropriate.

The architect and the law

THE LAW

Sources of Law

The United States' judicial system developed originally from English common law, and is aimed at preserving the fabric of society. It is embodied in:

- Federal and state constitutions
- Statutes
- Common law
- Regulations of administrative agencies

In addition, equitable doctrines, which allow for flexibility in decision making, are sometimes invoked.

Federal and State Constitutions

The US Constitution represents the supreme law of the nation, laying down rules which bind all aspects of government. Much of its content, notably the Bill of Rights, derives from concepts which emerged through the common law.

The Constitution is the highest source of US law and neither judge nor legislature may ignore or contravene its principles. Within the Constitution, however, the states have authority delegated to them to regulate public health, safety, and welfare in the form of building codes and regulations.

In addition, individual states have their own constitutions which are largely based upon the national model.

Statutes

Statutes are written laws officially passed by federal and state legislatures. Federal laws apply nationally, whereas state laws are only relevant to the state in which they are passed, and can vary throughout the country on the same subject (for example, professional licensure).

Common Law

The basic "rules" of society have emerged through the common law which demands that judges decide each new case on the basis of past decisions of the superior court. The principle of *stare decisis* (to stand by past decisions) is not a completely rigid concept: a judge may distinguish a new case from its predecessors in certain circumstances, thereby creating a new precedent. This enables the common law to grow and adapt according to the changing values and needs of society.

Where a conflict arises between a common law decision and a statute, the latter always prevails. Often an undesirable common law rule is disposed of by the passing of a statute.

Regulations of Administrative Agencies

Administrative agencies are often empowered to make and enforce regulations which have the force of law.

Equity

The concept of equity allows for additional procedures and remedies to be granted in court proceedings. It provides a measure of fairness not always available under rigid statute or common law. For example, if an owner avoids payment on the basis of a legitimate contractual technicality, the architect might claim based on the principle of unjust enrichment.

Classification of Law

Law pertaining to the practice of architecture can be classified into four basic categories:

1. Criminal law
2. Civil law
3. Civil rights law
4. Administrative law

Criminal Law

Acts committed against society or the public good by individuals which are proscribed by federal or state laws are generally classified as crimes (e.g., murder, theft, etc.). Lesser crimes are called misdemeanors, whereas more serious offenses are known as felonies. Some states prohibit professional licensing for individuals with a criminal record.

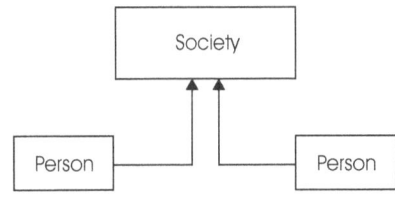

Figure 1.1

Civil Law

Civil law is private law dealing with the rights and obligations of individuals and corporations in their dealings with each other. Areas covered under this category include:

- Succession
- Family Law
- Contract
- Property
- Tort

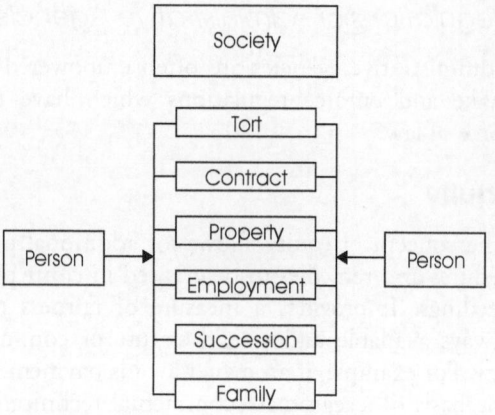

Figure 1.2

For the professional design practitioner, the most relevant branches of civil law are:

a. Contract law
b. Tort

Contract Law   This concerns the legally binding rights and obligations of parties who have made an agreement for a specific purpose (see page 63).

Tort   A tort is literally a "wrong" done by one individual (or corporation) to another for which a remedy (e.g., compensation, injunction, etc.) may be sought in the courts. Examples of specific torts are:

• Negligence (see page 6)
• Trespass (see page 50)
• Nuisance (see page 50)
• Defamation (see page 28)

It is possible for a case to fall under both contract and tort simultaneously (for example, where a negligent act results in a breach of contract). In these circumstances, it is often easier to sue on the contract rather than attempt to prove the tort.

Civil Rights Law

Civil rights legislation, such as the Americans with Disabilities Act, protects individuals against discrimination based on physical disability. Specific design guidelines and regulations ensure access to public accommodation. Federal fair housing statutes and some state legislation ensure the accessibility to, and adaptability of, certain types of housing.

Administrative Law

Legislation at the federal, state and local levels establishes and enhances building codes and regulations. These are designed to protect the health, safety, and welfare of the public. Architects may be held liable for their violation, which may possibly affect their licenses.

THE COURTS

The United States has two hierarchies of courts:

1. Federal
2. State

At the head of both hierarchies is the US Supreme Court.

Federal Courts

Cases are heard in federal courts when a federal question is involved or when a dispute arises between parties from different states. In many cases federal jurisdiction is concurrent with state jurisdiction, but in certain matters the federal courts have exclusive authority. Examples include:

• Patent and copyright
• Actions in which the US Government is a party
• Cases involving federal criminal statutes

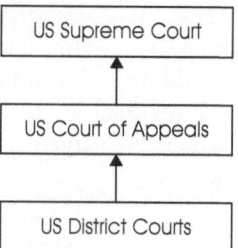

Figure 1.3

Federal trial courts are located throughout the United States. Each case begins at the district level, with the possibility of appeal to the relevant Court of Appeals and finally to the US Supreme Court. Criminal and civil matters are heard in all federal courts, although certain specialized courts exist for specific issues (examples include the Court of Claims, Court of Customs and Patent Appeals).

State Courts

State courts are limited in jurisdiction according to their location and the type of case involved. Generally, each state has at least two levels of trial courts. Criminal matters are heard at all levels, but frequently the lowest state courts are only authorized to deal with misdemeanors.

Similarly, civil cases are heard throughout the system, but the lower courts are restricted in their jurisdiction, often on the basis of the financial amount claimed.

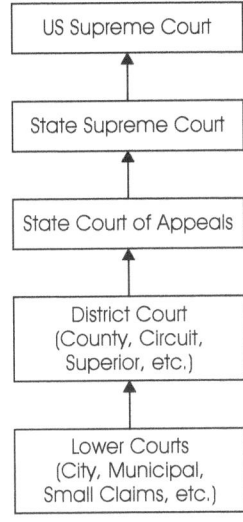

Figure 1.4

State court systems generally have two levels of appeals courts: intermediate courts of appeals and the State Supreme Courts. The final court of appeal is the US Supreme Court.

Small Claims Court

In many states, simple procedures have been developed for individuals wishing to sue for small amounts which would not be financially practicable in the regular courts system. The financial

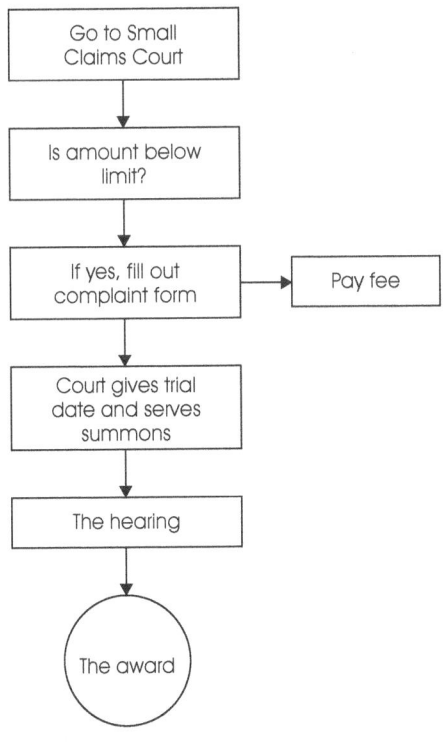

Figure 1.5

limit for small claims varies from state to state (but $5,000 is a common figure). In some states, professional representation is prohibited in these courts.

The United States Supreme Court

The US Supreme Court has original jurisdiction in cases involving disputes between states. In addition, it is the final court of appeal, but it will only hear cases it considers to be significant and which have originated in the state or federal courts.

Out-of-State Claims

Owing to federal due process requirements, some matters may be complicated if the parties are resident in different states. Many states have enacted *long-arm statutes* to enable suits to be brought against defendants resident in other states.

Standard of Proof

When a matter is decided in the courts, allegations must be proved. The standard of proof in criminal proceedings is very high: the prosecution must prove its case against the accused "beyond a reasonable doubt." In civil matters, parties need only prove their allegations to the degree that the court will accept them on a "balance of probabilities."

Other methods are available for the resolution of disputes outside the courts:

- Arbitration (see page 116)
- Mediation (see page 122)
- Administrative boards, agencies, and commissions (quasi-judicial forums which tend to be less formal than the regular courts and specialized in nature).

In most legal matters affecting design practice, it is advisable to obtain professional legal advice. Selection of an attorney may be facilitated by contacting a local or state bar association which, in many areas, operate convenient lawyer referral services free of charge.

THE ARCHITECT'S LIABILITY

The architect's legal obligations and responsibilities are owed to a variety of parties, and are governed by statutes, administrative regulations, and common law.

However, the majority of suits against architects are concerned with:

1. Breach of contract
2. Negligence

Breach of Contract

The architect enters into a contractual relationship with the owner to perform specific services (see page 36). An implied agreement is made by the architect to carry out the required work to the standards expected of the profession. Failure to meet these standards, which cause extra expense or delays for the owner, may result in a claim for damages against the architect on the grounds of breach of contract.

Negligence

Separate from any contractual obligations which may have been agreed upon, a duty or standard of care under the law of tort may exist (see page 4). If a person fails in this duty, a negligence suit could succeed. So the architect could be liable for the consequences arising from negligent behavior even in the absence of a contractual relationship.

The extent to which any party may be held liable to others in tort depends upon their specific duty or standard of care. In contractual situations, the obligations of both parties are usually clearly defined, but in tort it is often difficult to determine the extent or even the existence of a duty of care. However, some duties of care have been defined by case law and/or statute. Two of particular concern to the architect are:

- Strict liability
- Vicarious liability

Strict Liability

In certain cases, liability may exist independently of wrongful intent or negligence. This concept is best illustrated by the English case of *Rylands* v. *Fletcher* (1868), in which water from a reservoir flooded a mineshaft on neighboring land and led to a successful claim for damages, although no negligence on the part of the reservoir owner was proved. The decision against the owner was made on the basis that he had kept on his land "something likely to do mischief" and that it had subsequently "escaped." This made him automatically, or strictly, liable for the consequences.

The concept of strict liability has relevance to practice, for example, in the specification of materials, where the architect may be held liable for requiring new products that subsequently fail (see page 60).

Vicarious Liability

In some circumstances, one party is responsible for the negligent acts of another without necessarily contributing to the negligence. This is referred to as "vicarious liability" and a common example is the employer's responsibility for the acts of employees in the course of their work. A related example is the architect's liability for the defective work of consultants (see page 21).

In all cases concerning claims based on negligent behavior, certain conditions must be proved by the plaintiff if the claim is to be successful:

a. That a duty of care was owed by the defendant to the plaintiff at the time of the incident complained of
b. That there was a breach of contract
c. That the plaintiff suffered loss or damage as a result of the breach

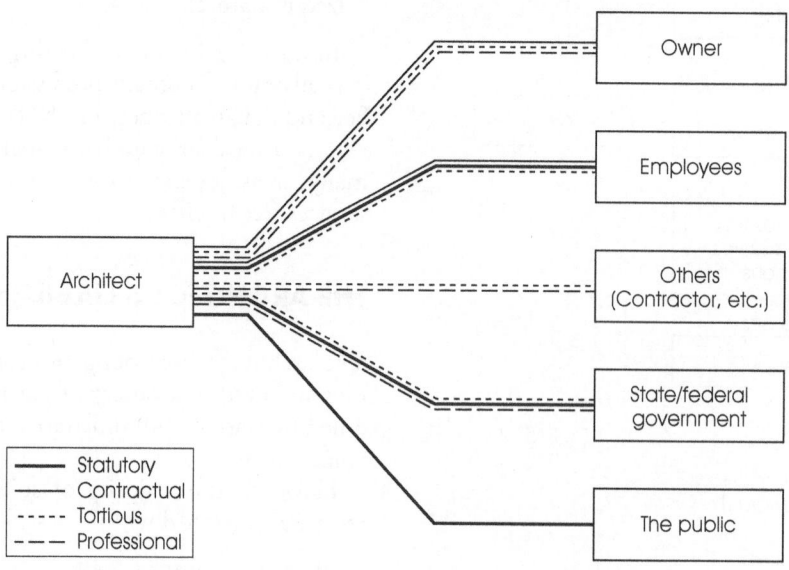

Figure 1.6

Standard of Care

In all cases, it is the "reasonable standard of care" established by common law against which a defendant's performance is matched and judged. In the case of the architect, the standard is considered to be the average standard of skill and care of those of ordinary competence in the architectural profession.

The Practice Overview on page 10 will give an indication of the extent to which an architect may be held liable for negligent acts, and also help to highlight the areas which merit particular care and attention. It should be noted that the architect's liability in tort is subject to periodic change as a result of changes in the law and, therefore, it is necessary to be constantly aware of new developments.

Criminal Liability

In certain limited cases, individual state law may impose criminal liability upon the architect (for example, if death results from the violation of a compulsory building regulation which expressly states that such a situation gives rise to a charge of manslaughter).

SAFEGUARDS AND REMEDIES

The law can be seen as a complex web of rules and procedures that enable and constrain the actions of individuals and groups. Breaking the rules,

whether intentionally or not, might lead to the implementation of prescribed punitive or compensatory measures.

In the construction field, a number of precautions and remedies are available to prevent or allow for certain contingencies. The most important of these are shown in Figure 1.7.

Insurance

Contracts of insurance may be entered into by the architect, the contractor, the subcontractor, and the owner to protect their respective interests. Under the AIA Document A201-1997 General Conditions (Article 11), provisions are made for owners and contractors to provide their respective insurance requirements with regard to property and safety and, optionally, project management liability.

Bonds

These fulfill a similar function to insurance: they enable an owner to claim relief from the surety who underwrites the contractor in the event of the latter's noncompliance with the contract requirements. Types of bond include performance bonds, bid bonds, and payment bonds (see page 74).

Warranties

These are assurances given by parties in respect of their goods and/or services (e.g., roofing) which

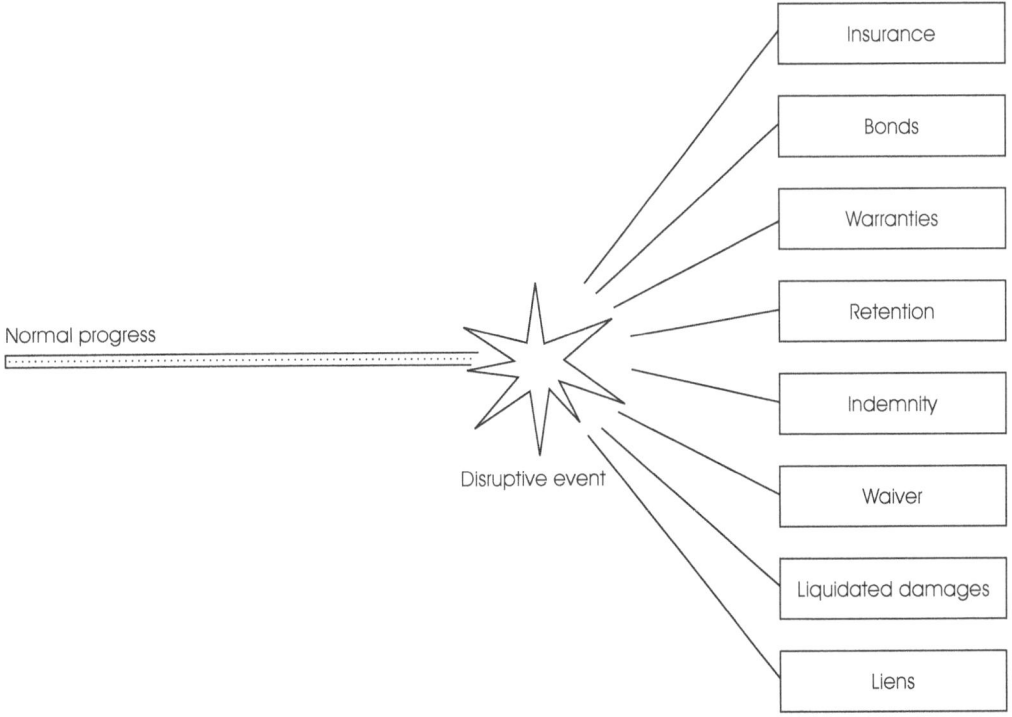

Figure 1.7

usually last for a stated period of time and are legally enforceable.

Retention

Before each progress payment is made during the construction phase, an agreed percentage will sometimes be retained by the owner to ensure the contractor's continued performance until the completion of the work, when the accumulated sum is released. Though a prudent precaution for owners, retentions are unpopular with contractors and, in recent years, retained amounts have tended to be increasingly lower.

Variations of the procedure include retaining a percentage for the first 50 percent of the work only, after which the retainage, with the consent of any surety, may be reduced or discontinued. Alternatively, an agreed percentage may be retained upon the first 50 percent on each line item of the work, enabling subcontractors to benefit from early release. Some parties may agree to invest the retainage in order to accrue interest payable to the contractor upon successful completion of the work.

Indemnity

One party may secure or "indemnify" another against liability for loss or damage resulting from certain circumstances (e.g., AIA A201, Article 3.18). Indemnity may be implied by events, but, in the construction industry, it is generally considered good practice to express it in a written contract. Legal actions against architects are frequently based on differing interpretations of implied indemnity.

Waiver

A waiver indicates the giving up by one party of rights which may prevail over others (for example, in some instances, the acceptance of payment may constitute the waiver of certain claims against the payer). Waiver of some rights is restricted by individual state laws (such as waiver of lien: see below).

Liquidated Damages

These represent a formula specified by the contract documents which provides an agreed method of assessing damages, arising from late completion (e.g., $x per day, to be paid by the contractor to the owner for every day by which the agreed completion date is exceeded: see page 92).

Liens

In cases where goods and/or services have been provided, the supplier may be able to secure a private mechanic's lien or "hold" upon the recipient's property to ensure payment of outstanding fees. The applicability of lien laws varies from state to state, particularly with regard to professional services. A lien effectively encumbers the title of the property and may be released after satisfactory settlement of the debt.

Some states allow the architect to impose a lien for design work and administering the contract, whereas other states only allow a lien for work done by the architect on site. A few states do not permit the architect any liens at all. In view of these considerable variations, individual state lien laws should be carefully noted before attempting to make use of this remedy.

Claims: Settle or Defend

If a claim is made upon the basis that legal obligations have not been fulfilled, the party so charged may admit responsibility and settle the claim by agreed damages or other appropriate means of compensation. Alternatively, the claim may be denied, in which case it is likely that the dispute will be resolved either by litigation (through the civil court system), arbitration (see page 116) or mediation (see page 122).

Shared Liability

It is possible that more than one party will be cited in a tort action on the basis that they share responsibility for the act or omission complained of. In these circumstances, the cited parties may become *joint tortfeasors*.

Time Limits

Lapse of time may affect the validity of a civil court action, and individual states have promulgated limitation statutes. These vary, not only as to the time limit for bringing an action, but also as to the commencement of the limitation period (see page 109).

INSURANCE

A contract of insurance is created when one party undertakes to make payments for the benefit of another if specified events should occur. The conditions upon which such a payment would be made are usually described in detail in the policy. The consideration (see page 63) necessary to validate the insurance contract is called the *premium*.

Types of Insurance

The most important types of insurance relating to the construction process are:

1. Professional liability
2. Public liability
3. Construction contract

Professional Liability

In the light of current statistics, which indicate a significant number of negligence suits against the architectural profession each year (see page 10), liability insurance is a valuable means of providing financial protection. However, there is no legal requirement to insure, and some architects prefer to risk the consequences and save the high cost of premiums. Some clients, however, may require proof of insurance as a prerequisite to employment.

Professional liability insurance (often referred to as E & O, or errors and omissions) varies from company to company both in coverage and conditions, and great care should be taken in policy selection. In particular, the time limitation on claims under the policy should be checked (to discover whether the policy covers errors made prior to the policy period, which only become apparent during the policy period). Joint ventures (see page 19) are not covered automatically by professional liability policies, and at the outset of a joint venture agreement the architect should contact the insurer to request the necessary coverage.

Even the most careful and experienced architect should consider the security afforded by professional liability insurance, particularly because:

a. even if not negligent, the architect must still finance the defense of claims, unless protected by a suitable policy;
b. the architect is vicariously liable for the errors and omissions of employees; many professional liability policies provide coverage against this contingency.

Public Liability

Most architects, whether or not insured under a professional liability policy, carry a comprehensive general liability policy to protect against claims involving injury to persons or damage to property in connection with the architect's business or premises. These policies often exclude the risks specifically covered by professional liability policies. In addition, the architect in practice may require:

Employee-related insurance:
- Workers' compensation
- Disability
- Medical
- Retirement
- Death/dismemberment
- Group life

Office-related insurance:
- Building
- Building contents
- Documents
- Business interruption
- Criminal loss
- Motor vehicles

Construction Contract Insurance

In most building contracts (e.g., Article 11 of AIA A201), both parties are required to insure against contingencies relating to personal injury and property damage resulting from operations on site and, optionally, project management protective liability.

Points to Remember

Advice by the architect to the owner on matters of insurance should be avoided and may be specifically prohibited in some professional liability policies. Similarly, many types of policy become voidable if the insured fails to follow instructions prohibiting admission of liability. Policies should be read carefully to avoid potentially expensive errors.

Contracts of insurance are said to be of "the utmost good faith" (*uberrimae fidei*). This means that all material facts which might affect the insurer's willingness to accept the risk must be disclosed. Failure to disclose may render the contract voidable (see page 63).

Insurers should be notified immediately of all events which may affect the policy (e.g., changes in personnel).

Regularly check that the amounts of coverage are adequate, bearing in mind inflation, new acquisitions, etc. Keep all policies in a safe place. Ensure that renewal dates and premium payment dates are carefully noted so that policies do not lapse through inadvertence. Never take insurance cover for granted. If in doubt as to whether a risk is covered, check with the insurers promptly and ask for confirmation of specific coverage in writing.

Although personally unconnected with construction-related insurance policies, the architect should ensure that evidence of insurance required from the contractor has been approved by the owner prior to any certifications for payments.

PRACTICE OVERVIEW

LEGAL LIABILITY IN PERSPECTIVE

Legal liability has been a long-standing concern for architects, but just how serious an issue is it for contemporary practice? A brief historical overview may help to bring perspective to both the extent of the problems faced by the profession and the nature of the risks involved.

During the 1970s and 1980s, it was not uncommon to hear that over one-third of practicing architects were likely to be sued each year.[1] Much of that information, however, tended to concentrate on *why* the situation had developed without too much attention being paid to *what* the threat was. In the absence of any reliable database clarifying and quantifying the nature of legal liability, it remained largely undefined and, as such, was all the more disturbing by its vagueness.

Today, liability is still prominent as a focus, although much has been achieved to both understand and alleviate the threat.[2] During the 1980s, significant strides were made in dealing with the types and sources of liability claims.

First, it appears that the early estimates of the incidence of legal action were relatively accurate. The AIA reports that in 1978, thirty-five claims per one hundred insured firms were reported by architects and that by 1984, this figure had risen to forty-four.[3] These figures, of course, do not take into consideration action taken against uninsured architects or claims that were settled without recourse to insurers. Fortunately, these alarming increases subsided throughout the 1990s and are now around twenty claims per hundred. Second, information concerning the nature of architects' liability has provided a clearer indication of the characteristics of each lawsuit, and has helped to identify the areas of greatest concern. Perhaps most interesting is the high proportion of claims generated by alleged errors in the design phase. Assumptions that the majority of cases arise from construction-related problems are at variance with a number of sources. For example, the AIA has estimated that 78 percent of property damage suits blame errors in the design and/or contract documents for building failure. A study undertaken in Colorado also found that the design phase was the major source of litigation:

> The projects sampled in this study experienced an overall additive claim rate of 6% (i.e., 6 cents on the dollar) and, furthermore, 72% of these increases were due to design error or owner initiated changes. The more volatile issues so prevalent in the literature (delay, differing site conditions, maladministration, etc.) account for only 28% of the claims.[4]

The combined findings of these sources tend to suggest that architects seeking guidance on litigation-free practice should pay more attention to aspects of design than may otherwise have been considered necessary.

In addition to this finding, the information highlights the danger areas where architects typically become involved. The cases indicate an expansion in liability over time not simply in the number of cases involving architects each year but in both the range of duties expected to be fulfilled and in the heightened expectation of the architect's performance. Areas of contention that have become more prominent include third-party claims, cost estimates, responsibility for shop drawings, and even slander, although perhaps the two areas that stand out most clearly both in the number of cases involved and in their serious implications to the profession are the limitation of liability and implied warranties. In the first, cases reported throughout the United States[5] have involved statutes of limitation and repose, which have been interpreted in some states to render the architect accountable for errors for a virtually limitless period of time. Even death appears to be no protection against these claims. In one of the more extreme cases, the decision to allow the liability period to commence when the fault was discovered (and not at an end of the construction period, as was generally held in the past) resulted in a claim *against the estate of a deceased architect*, the residue of which was providing security for his widow.[6] Fortunately, many states have sought to limit the potential of never-ending liability through the enactment of "long-stop" statutes (a longer period of time during which claims may be brought but starting on a specified date).

The question of warranties, or the degree to which architects should be expected to guarantee their work, also raises some concerns. Strict, or automatic, liability has yet to be completely successful in arguments against architects in the courts. Nevertheless, decisions in the field of product liability have been used to suggest that complete building elements, such as roofs, are in fact products, and as such should render their designer strictly liable for their performance. These expansions of the architect's duty, in this case to a point where no fault needs to be proven to attach liability, is reflected in a number of cases, and suggests that the difference between a warranty and satisfactory performance is becoming less apparent. Two cases are illustrative of the high standards expected of the architect. Both seem ridiculous in their claims, and in fact both were decided in favor of the architects (who, of course, still had to pay legal fees and may have lost their deductibles).

The first case, brought against an architectural firm for negligent design of a prison facility, was instigated by the family of a prisoner who had committed suicide in his cell. The plaintiffs claimed that the architects should have designed the cells in such a way as to preclude the likelihood of self-inflicted damage. In the second case,[7] a zoo employee was injured while feeding an elephant, and sued the architect for failure to design the cage properly.

Both cases, although seemingly frivolous, were considered to be sufficiently substantial to make an adequate case against the architects' failure to exercise reasonable care in the designs. Although these cases failed, similar ones in the past, which at the time seemed unlikely to succeed, were successfully brought against the architects, increasing the standard of care for the profession as a whole. Such cases tend to highlight the boundaries of "safe" practice for the present, while indicating new areas of concern for the future and bringing the concept of implied warranty closer to reality.

Given the high level of legal liability, what has the impact been on the profession in real terms? Apart from general anxiety engendered by involvement in legal action and potential loss of reputation, the most dramatic, quantifiable impact can be calculated in insurance rates. Although it is a relatively new phenomenon (errors and omissions insurance became available in the United States only in 1956, although policies were drafted by Lloyd's of London soon after World War II), insurance costs have risen to the point where an annual premium has accounted for as much as 4 percent of the gross income of a practice, second only to payroll as a practice expense.

It has been suggested that at least part of the increased cost should be passed on to the client. In a highly competitive and expanding profession, however, firms may not want to risk losing work by increasing their fees. The result may lead to lower wages and reduced profit.

Is the current liability situation a serious problem for the practicing architect? There are some signs of encouragement and hope. For example, national insurance figures suggest that more than half of claims are settled without payment to the plaintiff, and that in two-thirds of the cases, the architects are victorious in court.

In addition to these figures, the increased understanding of the liability threat has raised the consciousness of the profession as a whole. This has led to the proliferation of guidance and warnings in the form of books, newsletters, articles, and workshop seminars, which are directed towards the self-protection of firms and the individual practitioner through understanding of the dangers and pitfalls involved in practice, and a commensurate lessening of malpractice claims.

Perhaps more significantly, liability has become a major issue at the professional level, and initiatives for reform in state legislation regarding liability, frivolous claims and tort has made some progress.

In conclusion, legal liability continues to be a sobering reality for the architect, although it is encouraging to see that the threat is now more clearly perceived and understood. In addition, action at both the individual practice and institutional levels has led to a more stable and secure future for the profession.

References

1. *New York Times*, 12 February 1978.
2. Dickmann, J.E., "Construction Claims—Frequency and Severity," *Journal of Construction Engineering and Management* 111, no. 1, March 1985 (a Colorado study), and Greenstreet, R., *Legal Impacts upon the Profession of Architecture: The Liability of the Architect in Wisconsin*, Center for Architectural and Urban Planning Research, University of Wisconsin-Milwaukee, 1985.
3. AIA Memo Newsletter of the American Institute of Architects, September 1985.
4. Dickmann, "Construction Claims."
5. Greenstreet, R., "The Limitation of Liability," *The Wisconsin Architect*, May 1985, 5.
6. Cecil, R., "Writing your Will to Defend your Estate from Eternal Liability," *Royal Institute of British Architects Journal*, December 1982.
7. *LaBombarbe* v. *Phillips Swager Associates*, 474 N.E. 2d 9 42 (Ill. App. 1985).

Question & Answer

Liability insurance can be very expensive and a number of practices I know opt not to carry a policy. Is this a wise idea?

Errors and omissions insurance can be expensive and has in the past cost as much as 4 percent of gross, an expense second only to payroll. While premiums depend upon the "hardness" of the insurance market, they have risen in recent years and some smaller practices have elected to "go bare." This strategy, which is risky, may be accompanied by the building of a "disaster" fund, essentially an investment of the premium amount in an interest-bearing account that may be used in the event of legal action. The advantages include a healthy saving of the accumulated premiums (if the practice remains litigation free) and a potentially lowered claims profile—an uninsured architect is probably less of a target, after all. The disadvantages are financial trauma if legal action occurs before an adequate pool can be saved and the likelihood of fewer clients, because many will require insurance coverage as a prerequisite for employment on anything but the smallest projects.

While insurance is not a universal panacea for protecting the architect against claims— there is usually a deductible and a limit to coverage—some of the national carriers provide a useful and often necessary component of successful practice and may offer extensive information, education, and training that can limit claims through improved practice.

The building industry

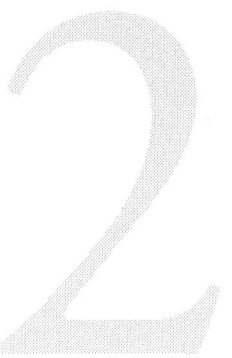

FORMS OF OWNERSHIP AND ASSOCIATION

Parties operating within the construction industry have different legal personalities according to their form of association. There are several methods of carrying on a business:

1. Sole practitioner
2. Partnership
3. Corporation
4. Joint venture
5. Other

Before setting up any type of business, it is advisable to obtain professional legal and financial advice.

Sole Practitioner

This is the simplest business form, with all liabilities and responsibilities vested in a single person. It is considered an appropriate organizational form for a small business with a predictable small-scale workload and a limited number of employees.

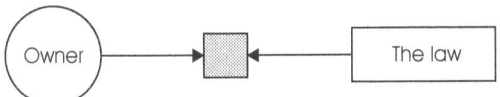

Figure 2.1

Partnership

A partnership exists where two or more individuals carry on a business as co-owners for profit. All profits are shared between the partners in previously agreed proportions. The Uniform Partnership Act has been adopted by most states, and it governs the major principles of partnership law.

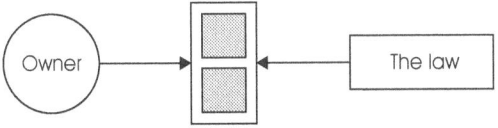

Figure 2.2

Partnership has become a common method of operating an architectural business as it enables architects to share their expertise, capital, and other resources.

The formation of a partnership does not limit the liability of individual partners, and each partner is responsible for all negligent acts and omissions of the firm jointly and severally, or in other words, whether personally negligent or not. However, partners joining the firm before, or leaving it after, a negligent act may be afforded protection.

In recent years, however, many states have created legislation that has allowed architects to practice as *limited liability partnerships*, where a partner is not necessarily personally liable for liabilities, debts, and obligations of the partnership other than for his or her own negligence, or that of someone acting under his or her control.

Formation

The partnership relationship can be created by:

- Conduct of the parties
- Oral agreement
- Written agreement

Most satisfactory is the written agreement, in which all aspects of the relationship can be expressed, thereby limiting the potential for disagreement or misunderstanding. In some states, all partners in architectural firms are required to be licensed architects.

Types of Partner

There are two major categories of partner:

1. The general partner
2. The limited partner

The General Partner   Unless otherwise arranged in the partnership agreement, all partners are deemed to have equal rights and liabilities within the firm, and all profits of the firm are divided equally in the absence of an agreed ratio. Similarly, all authorized acts of the partners bind the partnership.

Some partnerships may agree to take junior partners into the firm. As the title suggests, junior partners have less authority and control of the business, and take correspondingly lower responsibilities (usually restricted to personal acts and omissions). Profit-sharing will also be limited at this level. Care should be taken by all prospective junior partners to ensure that their position is clearly and accurately described in the written agreement. Further attention should be given to dealing with the public so as to avoid a general assumption of equality, and therefore joint liability, with the senior partners (for example, letterheads should be clearly marked, indicating the junior partner's name and position, distinct from those of the senior partners).

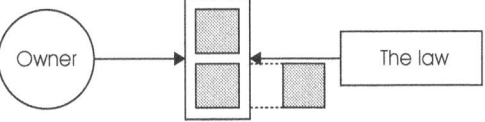

Figure 2.3

The Limited Partner Limited partners may invest capital in a firm and share profits, but they cannot be involved in the management of the business. Unlike general partners, their liability may be restricted to the extent of their investment. Limited partners are allowed in most states under the Uniform Limited Partnership Act, but they are not common in architectural practices.

Termination of Partnership

The partnership agreement may be terminated by:

- Expiration of an agreed time period
- Completion of a designated project or task
- Death of a partner
- Bankruptcy
- Retirement of a partner
- Mutual agreement
- Court order
- Subsequent illegality (see page 63)

Taxation

Partnerships are not taxed as distinct entities, and all partners pay individual tax upon their share of the partnership profits. Consequently, larger organizations may prefer to become incorporated in order to take advantage of tax concessions often available to corporations.

Partnership Agreement Checklist

- Date of agreement and names and signatures of the parties
- Date of termination (if any)
- Name and purpose of partnership, and business address
- Contribution of capital, provision for withdrawal, interest on capital, etc.
- Division of responsibilities and duties within the firm
- Salaries and profit-sharing details
- Methods of accounting, banking, etc., including specification of the partnership's fiscal year
- Insurance
- Benefit schemes, including pensions for outgoing partners and their families
- Rights of all partners in case of death, sickness, retirement, and withdrawal
- Arbitration/mediation agreement
- Length of vacations
- Provisions for check-writing
- Provisions for hiring and firing
- Procedure relating to loans by partners to the partnership
- Provisions in case of disqualification, bankruptcy or misconduct of a partner

- General provisions for dissolution
- Admission of new partners

The above checklist is by no means exhaustive, and architects should note that the more detailed and specific the partnership agreement, the less chance for future problems.

Corporations

Corporations are legal entities suited mostly to larger scale operations, and owned by (although distinct from) their shareholders. Corporations can be characterized by:

- perpetual existence of independent, individual shareholders;
- profit-sharing by shareholders;
- limitation of liability of shareholders to the extent of the value of their personal share obligation (except in limited circumstances where the so-called "corporate veil" can be pierced by a court to enable an injured party to seek redress).

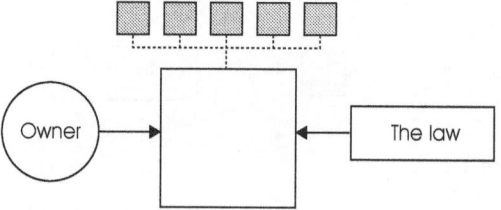

Figure 2.4

All corporations are subject to the law of the state in which they are incorporated. In addition, each corporation has its own Articles of Incorporation which generally draw the parameters of its activities, its organizational structure, and shareholders' rights.

There are three major types of corporation:

- Profit corporations
- Nonprofit corporations (e.g., charities)
- Professional corporations

An architect may generally be a shareholder in a corporation as long as it does not affect his or her professional duties. In recent years, many states have enacted statutes to enable architects to set up *professional corporations* in which to practice architecture.

Professional Corporations

Professional corporations differ from other corporations in that, although liability can be limited in certain contractual matters, the individual professional remains personally responsible for all negligent acts or omissions despite the incorporation.

Consequently, an errors and omissions (E & O) insurance policy is advisable for architects who are members of professional corporations.

In some states, architects who practice in a professional corporation can avoid liability where the negligent act was totally outside their personal control. Individual state laws should be consulted to ascertain the position of members of professional corporations with regard to personal liability.

Major advantages for the architect in forming a professional corporation include certain taxation benefits, perpetual existence of the corporation, and limited security of personal assets. However,

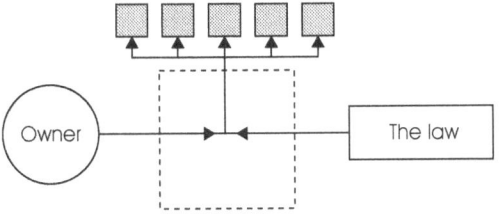

Figure 2.5

this form of association also has disadvantages such as administrative costs and formalities. Also, some public authorities may be unable to deal with professional corporations, and out-of-state work might be made difficult. For a variety of reasons, professional legal and financial advice should be sought prior to setting up a professional corporation.

Limited Liability Companies (LLCs)

While having many of the characteristics of companies, LLCs are taxed by the federal authorities as partnerships. State law varies, although typically architects in LLCs can limit their liabilities for acts or omissions not directly under their control.

Joint Ventures

If two or more organizations wish to combine forces for a specific project, they may engage in a joint venture. This is a type of partnership limited to the duration of the task. Advantages include:

- Shared resources
- Combined expertise and knowledge
- Joint capital
- Fluidity of staff allocation

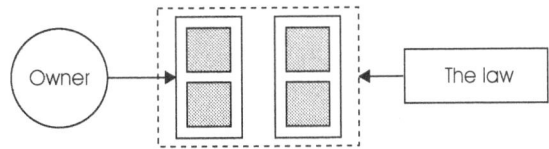

Figure 2.6

The arrangement must be conceived as a limited one, or it may be viewed by the taxation authorities as taxable on a corporate basis. If a joint venture is felt to be an appropriate means of temporary practice, the form of agreement between the organizations concerned should be carefully drafted, specifying the precise purpose of the venture, respective tasks and responsibilities, and compensation, using the same guidelines as those for a partnership agreement (see page 18).

Formation

There are two basic types of joint ventures:

- Fully integrated self-supporting joint venture
- Nonintegrated joint venture

The fully integrated self-supporting joint venture is formed when the organizations concerned create an entirely new association, separate from the original firms, which operates independently with a separate work force, payroll etc.

The nonintegrated joint venture is less formal and allows employees in each firm to undertake the work while remaining in their respective offices, and on the original firm's payroll. This is the more usual form of architectural joint venture.

Compensation

Firms engaged in a joint venture may divide the compensation from the venture in one of two ways:

a. Profit split
b. Compensation split

Profit Split By this method, compensation received from the owner is placed in a joint account and divided between the venturers (after expenses have been deducted) according to an agreed formula.

Compensation Split This method allots a portion of the project's compensation to each venturer at the outset, and then offsets the costs of the services necessary to complete the work against the sum allotted so that the difference is retained as profit. This means that firms which operate efficiently avoid financial loss caused by the inefficiency of other firms.

In some circumstances, architects will form joint ventures with a view to being commissioned for a particular project. Rather than undergo the full requirements before the work is assured, the details of the proposed venture may be written down in a *memorandum of understanding*. This memorandum could form the basis of a full joint venture agreement if the firms are granted the commission.

Insurance can be taken out under each firm's existing policies with an appropriate endorsement, or by a separate policy in the name of the joint venture.

Other Associations

Other forms of organization which may be encountered in the construction industry include:

a. Associated architects, or "loose groups"
b. Professional associations and unincorporated associations
c. Trade unions
d. Governmental agencies (federal and state)

Associated Architects

The term "associated" with regard to architectural practice is vague, and may refer, among other things, to independent organizations sharing facilities, or to a nonintegrated joint venture of firms. The AIA recommends that the use of the term "associated" should be avoided unless the actual legal relationship of the parties is clearly defined. In the absence of a clearly defined relationship, a partnership may be implied by the courts, leading to complex and expensive liability problems.

Two forms of association are increasingly common:

1. Often a "design architect" works with an "architect of record" on specific design projects. The former establishes the conceptual and schematic basis for the project, while the latter takes responsibility for construction documentation and construction administration.
2. In large or complex projects, an "executive architect" may manage and coordinate the work of a "consulting architect" who is responsible for specific portions of the project.

Professional Associations and Unincorporated Associations

The professional association is not technically a corporation, but is sufficiently corporate to be treated as such for taxation purposes. Unincorporated associations (e.g., social clubs) are not legal entities, but in most states they do have limited legal capacity (e.g., to contract). Architects working for such groups should be careful to check the authority and liability of the members they deal with; this information can usually be found in the constitution or regulations of the association. State laws regarding the legal capacity of these associations should also be checked by

the architect before entering into a contractual relationship.

Trade Unions

These are groups formed within the trade (often as unincorporated associations) for the purpose of collectively bargaining for pay and conditions of employment.

Government Agencies

The regulations of these bodies, both at state and federal level, derives from statutes. They have, in the past, enjoyed immunity from legal actions. However, this immunity is now less absolute in many states, and a number of claims have been made successfully against governmental agencies for their negligent acts or omissions (e.g., negligent plan inspection).

THE PARTIES INVOLVED

Professional Relationships

The Architect/Owner

The relationship between the architect and the owner is primarily contractual, and as such is governed by the terms of the contract between them. The contract formalizes a relationship of agency in which the architect (the agent) acts as the representative of the owner (the principal), working solely in the latter's best interests.

Agents are expected to work with the level of skill normally associated with their profession or occupation, and to be concerned to prevent any conflict arising between their own interests and those of their principal. The agency authority of the architect is limited by the terms of the appointment, and the architect should be careful to avoid overstepping his or her authority. For example, ordering the contractor to undertake work where the latter acts upon the apparent rather than actual authority of the architect may constitute a breach of the architect/owner agreement. Should the owner wish to extend the powers of the architect beyond those specified in the signed contract to enable the undertaking of specific tasks outside the scope of authority, written authorization should be obtained by the architect before carrying out such work.

The agency relationship between the owner and the architect is not a general one, and the architect may act as the owner's representative only in areas specifically stated in the contract between them. Where a decision is needed on a question in which the agent does not have authority, the

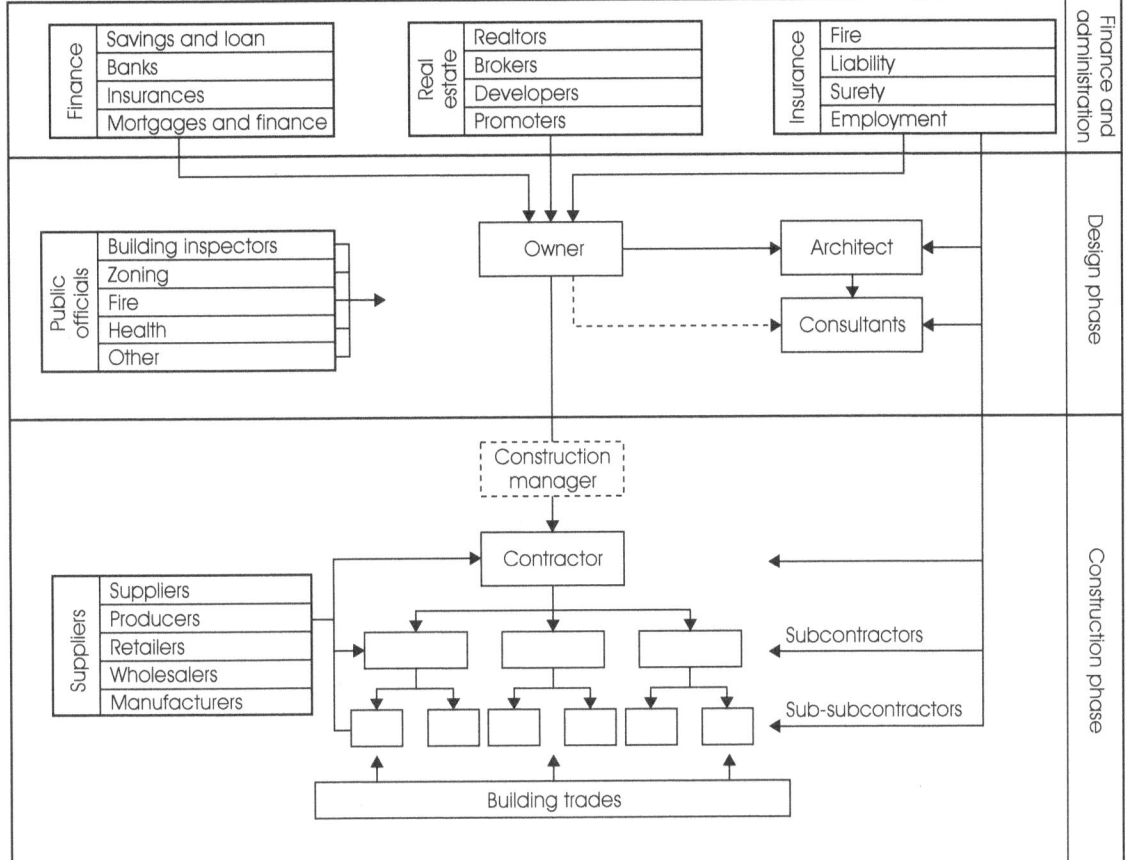

Figure 2.7

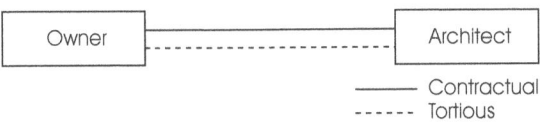

Figure 2.8

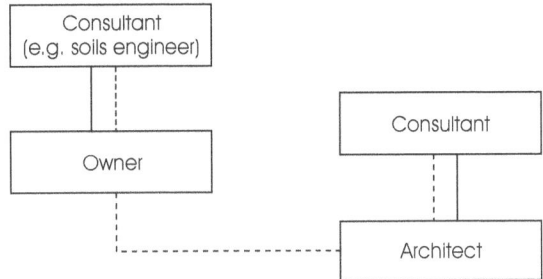

Figure 2.9

principal should be contacted. In an emergency, where the principal is not available, the agent is authorized to do anything which prevents loss to the principal. Such situations may give rise to dispute, and should be treated with the utmost caution.

Under the AIA A201 Contract for Construction 1997, the architect takes on a secondary role of quasi-arbitrator of the agreement between the owner and the contractor. Absolute fairness should be exercised in this role and, in spite of being the owner's agent, the architect must not show undue favor to the owner in the event of a dispute concerning the contract (A201. 4.2.12).

The Architect/Consultant

Where services necessary to a construction project are outside the architect's purview, specialists may be employed by either the architect or the owner

to undertake the work. It is usual for the architect to form a contractual relationship with a consultant although, in some instances, it may be possible for the owner to contract directly with the specialist (e.g., soils engineer).

Types of Consultant Consultants may be employed:

- for their technical knowledge (e.g., lighting, acoustics, landscaping);
- for their knowledge of specific building types (e.g., hospitals, theaters, schools);
- for other attributes relevant to a specific project (e.g., financial expertise, behavioral studies).

Care should be taken when employing consultants not to use their services for work which may fall under the architect's purview, as this may result in reduction of the architect's fee.

Selection   As the architect is vicariously responsible for the errors and omissions of the consultants, selection should be made with great care. Owner's recommendations may be considered, but the final choice should remain with the architect, who can and should require all consultants to maintain errors and omissions insurance coverage.

In order to fully delineate responsibilities, duties, and conditions of the relationship between the architect and the consultant, a written contract is advisable. The AIA produces two standard forms which are recommended:

- AIA Document C141, Standard Form of Agreement between Architect and Engineer
- AIA Document C431, Standard Form of Agreement between Architect and Consultant for other than Normal Engineering Services.

These documents are written to correspond with other AIA contracts (e.g., B141, A201, etc.) in terms of timing, format, and sequence. If a consultant's services are employed, the architect may be entitled to further payment to cover administration and extra risk. In some cases, the extent of work to be undertaken by a consultant may make it appropriate for the parties to engage in a joint venture (see page 19).

For limited or clearly defined work, a carefully drafted letter may serve instead of the full contractual documents. The letter should be sent to the consultant in duplicate with instructions to return one copy signed to the architect, and it should include:

- The names of the parties
- Date of the agreement
- Title and location of the project
- Description of the work
- Terms and conditions of service
- Payment type, method, and amount
- Insurance details

The Architect/Contractor

In conventional project delivery, there is no contractual relationship between the architect and the contractor, as the latter contracts directly with the owner. However, most building contracts contain provisions enabling the architect to undertake prescribed duties in the capacity of the owner's agent (see page 85).

Errors made by the architect which cause loss to the contractor could not result in an action under contract law (see page 63), but could form the basis for a claim against the owner who remains responsible for the agent's authorized

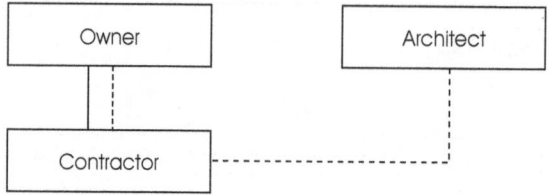

Figure 2.10

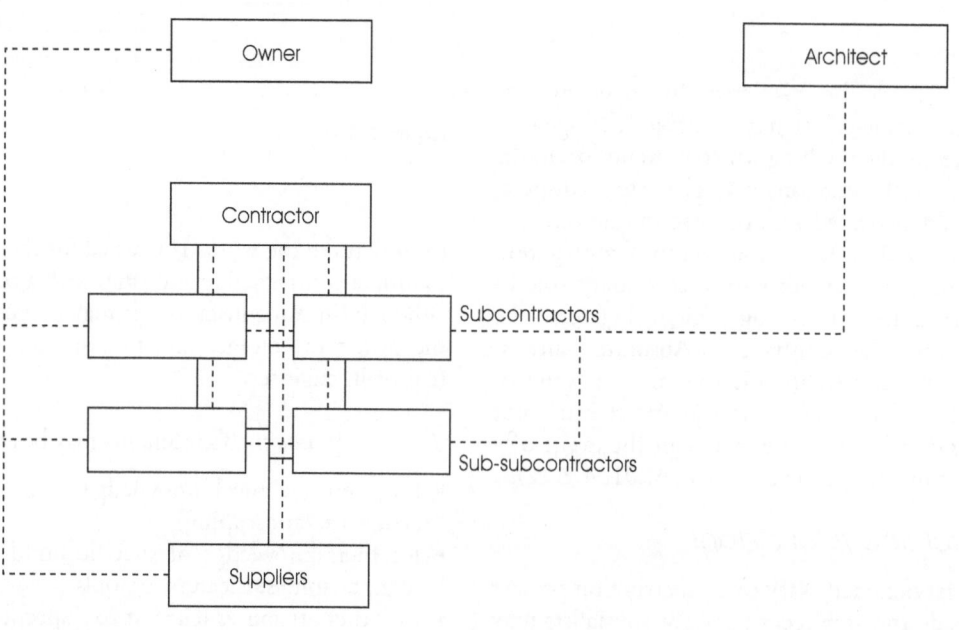

Figure 2.11

acts. This may in turn lead to an action by the owner against the architect for breach of the contract between them. Alternatively, the contractor could sue the architect in tort, where no contractual relationship is necessary (see page 6).

The same situation arises between the architect and subcontractors whose contracts are with the contractor, and also the suppliers who deal directly with the contractor and subcontractor.

The Engineer and Construction Manager

The Engineer

As in the profession of architecture, engineering work and the title "engineer" are usually protected under state law, although often the boundary between architecture and engineering work is ill-defined. In some states, engineers may be allowed to undertake work which might be considered to be architectural elsewhere, in addition to work primarily classified as engineering.

In any event, the professional engineer will normally be expected to conform to the examination, registration, and professional requirements of the state of residence, and will be subject to many of the practice-associated conditions which may apply to architects. The term "engineer" is a general description of many distinct fields of expertise, several of which are represented by their own professional bodies (e.g., the American Society of Civil Engineering). Engineering fields include:

* Soils
* Structural
* Mechanical
* Electrical
* Acoustic
* Highways
* Civil
* Drainage

Architects and Engineers Where architectural firms wish to engage the services of an engineer, it is advisable to use AIA Document C141, Standard Form of Agreement between Architect and Consultant. It is important to define the engineer's services as fully as possible in the contractual agreement, so that relative duties and liabilities can be determined and insurance coverage maintained accordingly. This is particularly relevant because, although the engineer must perform to the standard expected of his or her profession, the architect is usually vicariously responsible for an engineer's negligent acts of omissions.

The Construction Manager

The use of construction managers is an increasingly common practice for large and/or complex building projects, though the scope and detail of operations carried out under this term varies. Construction management services may be practiced by a number of parties. Some general contracting companies have entered the field, either in addition to or instead of normal construction work. Also, architects, engineers, and others with expertise and experience in the construction industry (e.g., construction superintendents) have undertaken similar services. The contractual arrangements made with a construction manager vary. Often, the contract is made directly with the owner, and the construction manager acts as go-between for all the parties involved in the building project and the owner. However, it is possible for such a manager to be employed as a consultant by the architect, or to form a joint venture with the architect (see page 19).

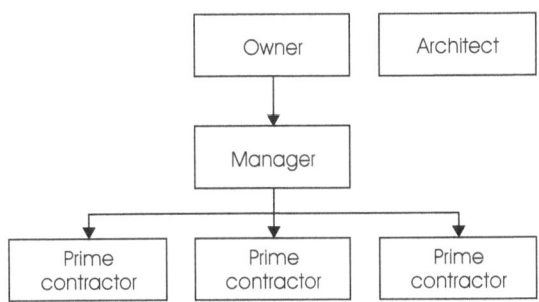

Figure 2.12

There are basically three primary roles undertaken by construction managers.

1. Advisers provide expertise on constructability, cost control, and construction methods. Advisers do not have a monetary interest in the means and methods of construction.
2. Agents organize and coordinate the various subcontractors and construction trades.
3. Constructors play an advisory role during design and then shift to the role of contractor for the construction phase. This dual role holds the potential for conflict of interest, as advice provided during design may unduly influence the overall cost of the project, and the constructor's profit.

AIA Document B144/ARCH-CM, Standard Form of Amendment for the Agreement Between the Owner and Architect where the Architect Provides Construction Management Services as an Adviser to the Owner, provides a means to integrate a construction manager role with that of an architect providing design and other construction administration services as described in AIA Document B141. Construction management carries with it a correspondingly high level of liability for actions related to supervision. Architects involved in construction management assume greater responsibility and authority during construction, but also face a correspondingly high level of liability.

Architects who offer services in this area should be careful to ensure that the scope of work and attached responsibilities are adequately defined in the contractual agreement, and that insurance coverage is correspondingly broad.

Other standard AIA documents that have been developed for use in these circumstances include:

- A101/CMa, Standard Form of Agreement between Owner and Contractor – Stipulated Sum, Construction Manager-Adviser Edition
- A121/Cmc, Standard Form of Agreement between Owner and Construction Manager where the Construction Manager is also the Constructor (AGC Doc. 565)
- A131/CMc, Standard Form of Agreement between Owner and Construction Manager where the Construction Manager is also the Constructor – Cost Plus a Fee, No Guarantee of Cost
- A201/CMa, General Conditions of the Contract for Construction, where the Construction Manager is not a Constructor, Construction Manager-Adviser Edition
- A311/CM, Performance Bond and Labor and Material Payment Bond, Construction Management Edition

- A511/CMa, Guide for Supplementary Conditions, Construction Manager-Adviser Edition
- B141/CM, Standard Form of Agreement between Owner and Architect, Construction Management Edition
- B141/CMa, Standard Form of Agreement between Owner and Architect where the Construction Manager is not a Constructor, Construction Manager-Adviser Edition
- B144/ARCH-CM, Standard Form of Amendment for the Agreement between Owner and Architect where the Architect Provides Construction Management Services as an Adviser to the Owner
- B801, Standard Form of Agreement between Owner and Construction Manager where the Construction Manager is not a Constructor
- G701/CM, Change Order, Construction Management Edition
- G701/CMa, Change Order, Construction Manager-Adviser Edition

The Design-Builder   Design-builders provide a one-stop source for design and construction services. The design-builder may provide all services or may subcontract parts of design services or construction work. Document A191 provides for flexibility in tax arrangements. Note that as with construction management, architects who participate in design-build may substantially increase their liabilities. Contractors are held to a vendor's standard of care which differs from that of the professional. The vendor's standard is based on performance of the work specified in the contract documents.

PRACTICE OVERVIEW

THE ARCHITECT/CLIENT RELATIONSHIP

The literature of architectural practice is relatively consistent in its treatment of the architect/client relationship. Whether the architect is conceived of as an expert adviser or likened to the supporting partner in a Victorian marriage,[1] the importance of establishing and maintaining close ties with the client is constantly stressed and, given the largely private, commercial nature of practice in the United States, may seem perfectly logical. However, research into the separate but associated areas of professionalism and legal liability suggests that a potential conflict may exist between the desire to maintain good architect/client relationships and the pressing need to guard against legal action.

As many architects are painfully aware, the threat of legal liability has escalated over the years, and every aspect of practice needs to be monitored closely to reduce the likelihood of court action. While the construction phase produces a significant share of legal actions, many cases originate from acts or omissions taking place during the design phase, where the architect and the client are the major participants.[2] Many of the problems in these cases originate from alleged errors in construction documentation, but a sizable number are concerned with conflicts between the architect and the client, not necessarily connected to design error. In a Wisconsin study,[3] for example, over one-fifth of the actions were initiated by architects suing for fees. In many of these cases, the architect/client relationship had broken down and refusal of further payment had precipitated the legal action.

Many law suits stem from inconsistencies and misunderstandings in client negotiations. In a number of instances, particularly in small-scale commissions for clients with little knowledge of or previous experience with the building process, architects had casual, informal contractual arrangements and a tendency to shield the client from potential construction problems. When problems *do* arise in such instances, the client is rudely awakened to his or her (often monetary) consequences and tends to blame the architect, not always unreasonably, for their occurrence. In order to minimize these pitfalls, the architect should follow more businesslike and formalized procedures that clearly define the rights and responsibilities of the two parties. However, by establishing clearly-defined legal boundaries between the parties, such actions seem to diminish or violate the architect/client relationship that has become accepted at the smaller scale of practice.

A review of common misunderstandings between the architect and the client suggests that they can be avoided in a way that does not violate or endanger the relationship. Closer attention to procedures both *before* a formal agreement is signed and during the contractual relationship may not only reduce the chances for legal action, but establish a sounder "professional" relationship which ultimately increases the likelihood of successful completion of the project.

At the beginning of the architect/client relationship, before contract formation, it is vital that the parties achieve a "meeting of minds," where the expectations and duties of both parties are clearly and unambiguously communicated and mutually agreed upon. This fundamental contractual principle may seem obvious to those practitioners with regularized procedures and broad negotiating experience. Nevertheless, many misunderstandings and disputes have arisen from situations that could have been easily avoided by clearer explanation and planning before the architect/client agreement was signed. For example, a number of cases suggest that, during early negotiations, some clients were not fully apprised of their responsibilities or the financial, legal, and temporal implications of the project. Potential problems were not mentioned, and even monetary matters were glossed over. It was not clear, for instance, how much the architect's fees would be, how they would be calculated, or even when they would be paid, so that the client's financial obligations to the project were not clearly articulated. Accordingly, when problems became apparent, the architect was often blamed for not alerting the client in advance. Whether these problems arose from insufficient briefing by the architect or by a misguided desire to shield the client from some of the rigors of the construction process, courts are likely to hold that, as the expert with a duty to "advise and consult" with the owner, the architect should bear responsibility for ensuring that full communication exists between the two parties, and that the client is aware of the obligations of *both*.

Architects approach client negotiations in different ways, and may not wish to risk losing potential commissions by appearing overly alarmist or pessimistic. However, it is more prudent to highlight the realities of the building process, with its complicated rules and procedures and attendant uncertainties, than to allow the client to discover these after the contract is well under way. Other aspects of the architect/client relationship that have proven problematic should also be discussed before signing the contract to ensure a clear understanding of the respective roles of the parties. It should be made clear, for example, that architects do not warrant perfection in their work, and that, despite every effort being made, some problems—in timing or cost, for example—may arise. Similarly, the extent of architectural duties that should be expected for compensation should be carefully explained using either B141 or B151 as a checklist. Problems have arisen where architects have either asked for payment after the fact for work that the client assumed was part of their basic fee (attendance at hearings, preparation of graphics or models for presentations, etc.) or where clients have been dismayed to discover a required task had *not* been fulfilled. A classic example of this lies in the expectation of detailed cost estimates instead of the more approximate preliminary estimates required in the AIA contract.

Once the contractual relationship has been formalized, both parties should have a clear idea of their respective roles. However, the rather flexible approach to contracts taken by some practitioners has led to problems that have found their way to the courts. If standardized documentation is used (and the AIA contract documents are likely to be the most representative of traditionally accepted practices), many potentially problematic issues will be underscored, particularly if such documents are used as a negotiation vehicle in the pre-contract phase. In this way, the architect can "educate" the client as to the expectations of the contract establishing the boundaries of the architect's duties and responsibilities. Where verbal or personalized contracts are used in preference to standardized documents, there is a greater risk of omission of information that might be a potential remedy should problems arise at a later stage. For example, few personalized contracts are likely to contain articles covering ownership of the drawings, yet it is possible that, in the absence of a specific statement such as that made in AIA document A201, the owner may be granted custody of the employee's work.[4] Similarly, mediation or arbitration procedures, which can offer a useful alternative to litigation, may be difficult or impossible to implement without prior written agreement of the parties. If standardized contracts are used other than AIA (e.g., client-generated documents), they should be checked very carefully to ensure that the architect's rights and responsibilities are not radically altered without the clear consent of both parties. It is possible that documents prepared by the client's attorneys may vigorously protect their client's rights at the expense of the architect.[5]

Adherence to and full knowledge of contractual obligations is as applicable to architects as to their clients. An exact understanding of required duties should ensure that adequate payment is received, that prescribed duties are undertaken and, implicitly, that the boundaries of permissible power are not exceeded. AIA contracts, for example, do not empower the architect to stop or change the construction work or to have any responsibility for safety on site. Consequently, any action by the architect beyond that required or permissible is both gratuitous and a potential focus for liability claims. Similarly, the architect should not give advice on bonding and insurance to the client: in fact, certain errors and omissions insurance plans specifically prohibit such advice and will withdraw coverage should such advice be given.

If standardized documentation is used, the architect should take advantage of *all* its provisions, which, research suggests, is not common. It is estimated, for instance, that of the architects who use AIA A201 (by no means the whole profession), *less than 5 percent* take advantage of the article requiring the contractor to submit a list of subcontractors' names, to which the architect may make reasonable objection.[6] Strict adherence to contract provisions not only minimizes problems that may arise, but provides an effective defense in the event of court action, if it can be shown that the firm performed its obligations under the contract in a reasonable and professional manner.

Despite the perception that more formalized, businesslike procedures conflict with the traditional architect/client relationship, the former are clearly desirable and not incompatible with the traditional image. More effective architect/client negotiations using standardized agreements will help minimize the misunderstandings and legal liability and even strengthen the relationship. The ultimate objective of the association is, after all, a project completed with the minimum number of disturbances or setbacks. Increased efforts by the architect to reduce possible conflicts can only serve to strengthen this goal and ultimately serve the expectations of the client in a responsible and professional way.

References

1. Maister, D., "Lessons in Client-Loving," *Architectural Technology*, Fall 1985.
2. In Dickman, J.E., "Construction Claims—Frequency and Severity," *Journal of Construction Engineering and Management* 111, no. 1, March 1985.
3. In Greenstreet, R., *Legal Impacts Upon the Profession of Architecture: The Liability of the Architect in Wisconsin*, Center for Architecture and Urban Planning Research, University of Wisconsin-Milwaukee, 1985.
4. Walker, N., Walker, F. and Rohdenberg, I., *Legal Pitfalls in Architecture, Engineering and Building Construction*, 2nd edn, McGraw Hill, 1979.
5. Greenstreet, R. "Who Really Owns Your Design?" *Progressive Architecture*, April 1985.
6. Kaskell, R., "How Do I Protect Myself from Suits by the Contractor?" in *Avoiding Liability in Architectural Design and Construction* (R. Cushman, ed.), John Wiley Interscience, 1983.

Question & Answer

Architects are getting sued for everything these days—I even hear you have to watch what you say or you could be sued for your opinions. Can this possibly be true?

The tort of negligence encompasses many aspects of the architect's activities which may include negligent misstatement. If professionals give expert advice which, if relied upon, leads to loss or injury, they may be held liable for the consequences. This responsibility, which does not require a contractual relationship or any form of compensation to be relevant, means architects should only offer their professional opinion with care. This includes the giving of references, any comments regarding safety or operations on site (which are not contractually required anyway) and advice on the selection of contractors, subcontractors and suppliers. In the last instance, there have been several cases where the architect has legitimately used contractual authority to reject subcontractors, only to be sued for defamation.

In any case, where a professional opinion or judgment is offered, it should be as objective, dispassionate, factual and accurate as possible. Any indication of malice, inaccuracy or bad faith, especially in a written form, could be the basis for a legitimate claim.

The architect in practice

THE PROFESSION

Each state enforces licensing laws which control the use of the title "architect" and the practice of architecture. Qualification requirements differ depending upon the standards of the state licensing boards, although the National Council of Architectural Registration Boards (NCARB), with support from the AIA, encourages uniformity in educational and examination procedures and facilitates reciprocity among the states. A combination of higher education, architectural experience under the supervision of a licensed architect, and passing a licensing examination, is generally required as a prerequisite to a license, together with other specified requirements. The state laws are enforced by regulatory boards.

Interstate Licensing

As the licensing of architects is carried out on a state-by-state basis, it may be necessary for the architect to requalify a number of times if engaged in work in several states. However, some states operate a reciprocity system to facilitate interstate practice, and enable architects to obtain a temporary or permanent license. NCARB certification greatly facilitates interstate licensing. Partnerships (see page 17) should ensure that all general partners are registered in states where work is undertaken. Failure to abide by state licensing laws renders the offender liable to imprisonment or fine, and may provide sufficient grounds for the owner to successfully avoid payment of fees.

Intern Development Program

Most states require an internship under the supervision of a registered architect as a prerequisite for license. NCARB administers the Intern Development Program (IDP) as a means to facilitate exposure of interns to a broad range of professional responsibilities and experiences. Working with a supervisor within the intern's firm and a separate mentor assigned to the intern, interns maintain detailed records of their experiences, and must attain a specific number of training units in order to satisfy licensing requirements.

ETHICS

The broader field of ethics is concerned with the norms, values and associations of individuals and groups, which may be embedded in cultural, religious or societal frameworks. They may be formalized in laws, or simply incorporated in accepted standards of behavior.

In a professional context, ethical behavior is often established in codes of conduct, which provide a collective assessment of required standards of professional performance with specific regard to society, to clients and to other members of the profession.

In the profession of architecture, codes are enforced at the level of registration within each state, and by the professional bodies that represent the field.

State Codes

Rules of behavior are enacted by each state and are codified in administrative regulations. They are mandatory, and violation may result in:

- suspension of license;
- censure;
- fine;
- imprisonment.

Many state codes are based upon the model developed by the National Council of Architectural Registration Boards.

National Council of Architectural Registration Boards Rules of Conduct

NCARB comprises the registration boards of all fifty states, the District of Columbia, Guam, the Northern Mariana Islands, Puerto Rico and US Virgin Islands. The Council has developed standards regulating architectural practice which are established in the Rules of Conduct.

The rules incorporate the following areas:

- Competence
- Conflict of interest
- Full disclosure
- Compliance with laws
- Professional conduct

The Rules of Conduct are recommended by NCARB to Member Boards and may be adopted in whole or part by individual state Architectural Registration Boards.

American Institute of Architects Code of Ethics and Professional Conduct 1997

The latest version of the code is a balance of mandatory and voluntary principles that conform to the Sherman Anti-Trust Act. It is administered by a National Judicial Council appointed by the AIA Board of Directors and is applicable only to AIA members. Licensed architects who are not AIA members are not subject to the provision of the code.

The code addresses responsibilities:

- to the public;
- to the clients;
- to the profession;
- to colleagues.

The code is arranged in three tiers:

- Canons
- Ethical Standards
- Rules of Conduct

Canons

These are broad principles of conduct which members are encouraged to meet.

For example, "Members should maintain and advance their knowledge of the art and science or architecture . . ." (Canon 1).

Ethical Standards

These are more specific goals to which members should aspire in professional performance and behavior.

For example, "Members should serve their clients in a timely and competent manner."

Rules of Conduct

The Rules of Conduct are mandatory and the only component of the Code of Ethics and Professional Conduct that are subject to enforcement.

For example, "Members shall not accept compensation for the services from more than one party on a project unless the circumstances are fully disclosed and agreed to by all interested parties."

THE OFFICE

Although architectural offices vary considerably in their structure, management, and workload, certain general observations and recommendations can be made regarding their administration.

Administration

Initial factors to be considered in the running of an architectural office include:

1. Insurance
2. Financial management
3. Office organization
4. Staff organization and selection

Insurance

As well as sufficient insurance to cover negligent performance by all employees for whom the principals remain vicariously responsible, insurance policies should be maintained to cover the office and its contents with respect to loss or damage, and also employee benefits (e.g., medical expenses, vehicular insurance, and compulsory coverage under the workers' compensation laws).

Financial Management

The necessity for maintaining accurate accounts cannot be too highly stressed, and professional assistance may be necessary to establish an accounting system and, possibly, to operate it. The AIA has prepared a manual for its members on the subject entitled *An Architect's Guide to Financial Management* (1997), by Lowell V. Getz, which architects in private practice may find of interest. The AIA Firm Survey periodically publishes billing data according to firm size and employees. A number of firms, such as Practice Management Associates and Zweig White & Associates, collect and publish updated surveys on financial performance, and accounting, fees, and compensation for architects.

Office Organization

To ensure efficiency and consistency within the office, it is advisable to record and maintain uniform procedures and techniques of office management. For example:

- Standardized communication methods (see page 85)
- Standardized drawing conventions
- Explicit roles, duties, and responsibilities for all personnel
- Use of standardized forms for office administration (see page 85). The AIA produces several of these including:
  G804 Register of Bid Documents
  G805 List of Subcontractors
  G806 Project Parameters Worksheet
  G807 Project Team Directory
  G809 Project Abstract
  G810 Transmittal Letter
  Other standardized paperwork may be used including accounting forms, fax sheets, telephone messages and memo pads, and order forms which can be printed in the house style.

In order to communicate and record information regarding office procedures in a consistent and readily available format, firms sometimes produce an "Office Standards Manual" containing the above data to provide a useful reference to new employees.

Staff Organization and Selection

Many practices consider it useful to establish office policy concerning their employees. A "Personnel

Policy Manual" is advised as a method of consolidating preferred practice both for existing members and prospective employees to familiarize them with office characteristics and expectations. The manual may contain general details of the practice (workload, direction, etc.), its organization, and fundamental policies regarding employment procedures and staff benefits. Information may include:

Office practice:
- Office hours
- Payment methods
- Overtime and time recording
- Lunch and coffee breaks
- Travel and expenses
- Responsibilities (equipment, etc.)
- Salaries and salary review
- Other concessions (parking space, etc.)
- Client service expectations

Staff benefits
- Pensions
- Profit-sharing and trust funds
- Holidays
- Incentive pay
- Sick leave
- Professional activities (further training, conferences, conventions, etc.)
- Civil duties (jury service, voting, civil projects, etc.)
- Dues to professional and civil organizations
- Future promotional policy

Hiring practices:
- Methods of selection
- Moving expenses and transfers
- Termination of duty, layoffs, and resignations
- Leaves of absence (military, emergency, etc.)

Contracts of Employment

A contract of employment need not be formulated in writing, but a carefully drafted agreement, covering all relevant issues established in the office manual will help to lessen the risk of future misunderstanding. Such agreements may be drafted into a letter of appointment, or attached to a letter in a standardized format, including the following details:

- Names of parties
- Date upon which employment commences
- Salary and payment intervals
- Hours of work
- Vacation period and vacation pay
- Sickness pay
- Pensions and other benefits
- Insurance coverage (professional indemnity, health, accident, etc.)

- Periods of notice
- Job title, duties, and responsibilities
- Required membership in professional associations
- Office benefits (e.g., automobiles)

Of course, professional legal advice is important in the drafting of contracts of employment. Finally, an architectural practice should be kept under continual review with regard to procedures, personnel, and equipment. In this way, timely adjustments can be made to assure the smooth running of the firm in the event of changed circumstances.

THE ARCHITECT/OWNER RELATIONSHIP

The architect/owner relationship will be affected by the nature of the owner, who may be:

- a private individual;
- a partnership;
- a corporation or institution;
- a state or federal department or governmental agency.

When dealing with employees of large organizations, it is advisable to check or verify their authority to bind the firm at the outset.

In some cases, the owner and the user of the proposed project may not be the same party (as, for example, in the case of a school or hospital), and care must be taken not to confuse the requirements of the two roles. Though users may seek features they deem appropriate, owners may not be willing to pay for them.

The character of the owner will affect the administration of the project in a number of ways, including:

- selection of the architect;
- the architect/owner agreement;
- contractor selection procedures;
- methods of construction;
- means of communication among the respective organizations;
- forms and paperwork to be used.

State and federal agencies are likely to want to use their own forms and contracts, and will be bound by statutory requirements in the selection of both architects and contractors.

Selection of the Architect

This may be accomplished in three ways:

1. **Directly**: Direct selection is a function of reputation, recommendation, previous contract, or chance.

2. **Comparatively**: This method is usually used by institutions, public agencies, etc., where a number of architects will be asked to submit their résumés for consideration by a board. Information required may include:

   a. age and achievements of the firm (examples of work, clever solutions, efficiency);
   b. details of the practice (staff, workload, organization, and ability to take on new work);
   c. references (bank, former clients);
   d. names of preferred consultants.

   Interviews may also form part of this selection method.

3. **Competitively**: Competitions may be:

   a. selective, where a limited number of entrants will be invited to participate;
   b. open, where anyone may enter.

   Architects are advised only to enter competitions approved by the American Institute of Architects, and abide by the guidelines it has established.

The Agreement

The form of agreement between the architect and the owner is very important, and care should be taken to clearly establish the relationship at the outset. Oversights, omissions, or misunderstandings at this stage may lead to serious problems later in the relationship which foresight and thorough attention could have helped to prevent.

There are a number of ways in which architect/owner associations can be formed:

- By conduct of the parties (see page 63)
- By letter (see page 63)
- By formal written agreement

A contract may be drafted for each new project, although the use of standardized forms is advisable. The AIA produces a number of standard forms although some owners, specifically larger institutions and governmental agencies, may wish to use their own standard forms. These should be studied carefully before signing as they may seek to increase the architect's liability beyond the standard of reasonable care.

In most cases, the use of AIA forms is strongly advised, as they are generally accepted and understood throughout the building industry and are comprehensive in their coverage. Less formal methods of agreement may be used in projects of a simple or minor nature where a full contract appears inappropriate. Here, abbreviated contracts may be useful (AIA Document B151) or carefully drafted letters of agreement, which might include:

- Details of the extent and purpose of the project
- The general nature of the agreement

- Details of the site (location and address)
- The responsibilities and roles of each party
- Payment type and times of payment (see page 38)
- Details of retainers
- Methods of calculating fees and expenses
- Details of full and partial services
- Copyright considerations (see page 38)
- Additional services, if any
- Other matters (consultants, type of building contracts, etc.)

Article Changes

In the eventuality that the standard contract articles have to be amended, omitted, or enlarged, great care should be taken to ensure that the terms of the agreement, as amended, do not adversely affect the architect's position with regard to liability, or conflict with provisions contained within related documents. If changes are necessary or required by the owner, legal counsel might be consulted to ensure that the overall documentation of the project remains consistent, and acceptable to the architect.

Checklist

Factors to be considered at preliminary meetings, and possibly mentioned in letters of agreement and/or contracts include:

1. Obtaining details of:
   - the owner and any representatives (names, addresses, etc.);
   - the project (description of intent);
   - the site;
   - the proposed user (if different from the owner).
2. Checking:
   - the seriousness of the owner and ability to proceed with the work (even a credit check of some clients might be prudent at this stage);
   - whether any other architects are involved with the project (if so, they should be informed);
   - the availability of office resources for the job;
   - statutory requirements and consents necessary.
3. Discussing:
   - appointment and payment of consultants;
   - type of building contract to be used;
   - method of contractor selection;
   - single or separate contract system (see page 66);
   - early appointment of contractor;
   - subcontractors and suppliers;
   - methods of insurance and security (bonds, warranties);
   - limitation of liability and indemnities.

1997 EDITION

AIA DOCUMENT | B141-1997

Standard Form of Agreement Between Owner and Architect with Standard Form of Architect's Services

AGREEMENT made as of the   day of
in the year
(In words, indicate day, month and year)

This document has important legal consequences. Consultation with an attorney is encouraged with respect to its completion or modification.

BETWEEN the Architect's client identified as the Owner:
(Name, address and other information)

TABLE OF ARTICLES

and the Architect:
(Name, address and other information)

For the following Project:
(Include detailed description of Project)

© 1997 AIA®
AIA DOCUMENT B141-1997
STANDARD FORM
AGREEMENT

The American Institute
of Architects
1735 New York Avenue, N.W.
Washington, D.C. 20006-5292

The Owner and Architect agree as follows.

1-1

AIA Document B141-1997: Standard Form of Agreement Between Owner and Architect

4. Providing the owner with:
   - details of the owner/architect agreement and information, including details of payment;
   - details of the architect to be in charge of the project;
   - methods of communication;
   - other data which will help the owner to understand respective duties and responsibilities and details of the construction process.

The Architect's Services during the Design Phase

While the architect and client can agree to any contractual agreement they like, it is always preferable to reduce the scope and details of the agreement to writing. Furthermore, the use of standard forms is highly recommended. These have been developed by all respective parties to the construction process and present a fair, balanced agreement that encompasses all the necessary details of a services agreement which may be encountered. There is also an industry-wide understanding of the meaning of each of the articles, which helps in any disagreements over interpretation.

Without a standard agreement, key issues affecting the relationship—arbitration procedures, ownership of drawings, etc.—may be unclear. Similarly, if a standard form is amended or supplemented, care should be taken not to adversely affect the rights of either of the parties involved, and legal counsel is advised should such an unnecessary step be taken.

American Institute of Architects Standard Forms

Clients may have their own contract, although the AIA produces a range of agreements which can be used:

- B141-1997 Standard Form of Agreement between Owner and Architect
- B141/CMa Owner-Architect Agreement, Construction Manager-Adviser edition
- B151-1999 Abbreviated Owner-Architect Agreement for Projects of Limited Scope
- B163 Owner-Architect Agreement for Designated Services
- B171 Interior Design Services Agreement
- B181 Owner-Architect Agreement for Limited Architectural Services for Housing Projects
- B727 Owner-Architect Agreement for Special Services
- B901 Design/Builder-Architect Agreement

The Architect-Owner Agreement 1997 Edition

There have been fourteen previous editions of the standard form beginning in 1917, although the latest 1997 format represents an important departure from its predecessors. The document recognizes and details the architect's duties. Rather than just defining the "basic services", it:

- attempts to clarify mutual responsibilities;
- quantifies respective roles;
- provides a mutual waiver of consequential damage;
- determines the loss of profit after the architect's termination.

The document is more flexible than previous editions, and provides benefits to both owners and architects, while creating a better overall agreement.

The 1997 Edition: Owner Benefits

- The architect must keep all information relative to the owner confidential.
- The architect must have no conflicts of interest.
- The architect must produce a project manual for every job.
- Criteria must be established for certificates, and better documentation must be kept.
- The architect will meet with the owner on final completion and in the post-construction period to review facility operation.
- If the project exceeds the original estimates, redesign will be undertaken by the architect without cost to the owner.

The 1997 Edition: Architect Benefits

- Better project parameters for compensation.
- Compulsory mediation.
- Contingencies can be included in estimates.
- Clearer involvement in the bidding phase and preparation of change orders.
- Ownership and use of drawings, including electronic materials.
- Better solutions for owner nonpayment.

The 1997 Edition: General Benefits

- The agreement (terms, conditions and compensation).
- The scope of services.
- Supplemental attachments.
- A more explicit services description and quantifications (e.g., number of site visits).
- Categorization of services by type, not phase.
- Linking compensation directly to services.

In summary, the 1997 edition tries to create closer communication between the architect and owner, promote careful negotiation (of both compensation and services) and thereby ensure mutual understanding of the agreement.

Scope of Services Replacing the original five phases of work (or basic services), the 1997 edition provides for sixty-eight possible services available to the owner divided into the following categories:

- Project administration services
- Planning and evaluation services
- Design services
- Construction procurement services
- Contract administration services
- Facility operation services

Each category contains a range of potential services that may be assembled and tailored to each individual project.

Project Administration Services

- Program management
- Management and administration
- Owner/consultant coordination
- Project presentation
- Special presentations
- Evaluation of project budget
- Schedule development and monitoring
- Preliminary estimate of cost of the work
- Detailed estimate of the cost of the work
- Owner-supplied data coordination
- Value analysis
- Agency consulting

Planning and Evaluation Services

- Programming
- Space schematics and flow diagrams
- Existing facilities surveys
- Economic feasibility studies
- Marketing studies
- Project financing
- Site analysis and selection
- Site development planning
- Detailed site utilization studies
- On-site utility studies
- Off-site utility studies
- Environmental studies and reports
- Energy studies and reports
- Zoning processing assistance
- Geotechnical engineering
- Site surveying

Design Services

- Architectural design
- Structural design
- Mechanical design
- Electrical design
- Civil design
- Interior design
- Landscape design
- Graphic design
- Special design
- Material research and specifications
- Special furnishings design

Construction Procurement Services

- Bidding/proposal documents
- Reproduction and distribution of bidding/proposal documents
- Special bidding/negotiation addenda
- Analysis of alternates/substitutions
- Pre-bid conference/selection interviews
- Bidding/negotiation
- Bid/proposal evaluation
- Contract award
- Report of bidding/negotiation results

Contract Administration Services

- General administration
- Submittal services
- Site visitation
- On-site project representation
- Payment certification
- Administration of testing and inspection
- Supplemental documentation
- Administration of changes in the work
- Interpretations and decisions
- Project close-out
- Construction management

Facility Operation Services

- Maintenance and operational programming
- Start-up assistance
- Record drawings
- Warranty review
- Facility operations and performance meeting
- Post-contract evaluation
- Tenant-related services
- Project promotion
- Leasing brochures

Additional Services

The agreement of services identifies a series of services that are not included in the general scope of services and would be provided only if specifically designated:

- Programming
- Land survey services
- Geotechnical services
- Space schematics/flow diagrams
- Existing facilities surveys
- Economic feasibility studies
- Site analysis and selection
- Environmental studies and reports
- Owner-supplied data coordination
- Schedule development and monitoring
- Civil design
- Landscape design
- Interior design
- Special bidding or negotiation
- Value analysis
- Detailed cost estimating
- On-site project representation
- Construction management
- Start-up assistance
- Record drawings
- Post-contract evaluation
- Tenant-related services

The Architect's Compensation

A range of methods is available to ascertain compensation for architectural services. The following are time based, reflecting the actual time spent on a project for which payment is calculated:

Multiple Direct Personnel Expense

The direct salaries of designated personnel are multiplied by a factor representing benefits, overhead, and profit.

Multiple of Direct Personnel Expense

The salaries of designated personnel are multiplied by a factor representing overhead, and profits.

Professional Fee Plus Expenses

The salaries, benefits, and overhead of personnel involved in the project represent the expense. The fee, or profit, may be agreed as a lump sum, a percentage or a multiplier.

Hourly Billing Rates

Salaries, benefits, overhead, and profit are included in the rate.

Other methods of calculating compensation are only indirectly tied to the actual time spent on the project:

Stipulated Sum

This is expressed in a finite, dollar amount.

Percentage of the Estimate of the Work

Compensation is calculated as a percentage of the estimated or actual cost of the work.

Multiple of Consultant's Billing

The bill of consultants hired by the architect is multiplied by a factor that accounts for the latter's administrative costs, overhead, and profit.

Square Footage

The overall square footage of the building is multiplied by an agreed pricing factor.

Unit Cost

Where appropriate, certain units, such as rooms, are multiplied by an agreed pricing factor.

Royalty

The architect's compensation may be calculated as a share of the owner's income or profit generated by the built project.

The compensation calculation methods may be used in differing situations or combined on the same project. Generally, the more uncertain the conditions—unresolved owner requirements, unusual site conditions, experimental—the harder it is for the architect to calculate the amount of time needed to accomplish the tasks. Accordingly, the time-based formulae are more appropriate. However, in some instances, such as publicly bid work, there may be no opportunity for negotiation and the payment type may be prescribed.

Reimbursable Expenses

In addition to the compensation method agreed upon, the architect may charge for expenses directly incurred during the project, including:

- Transportation connected with the project
- Fees paid for securing approvals
- Reproductions, documents, postage, etc.
- Overtime, if approved in advance by the owner
- Renderings, models, and mockups requested by the owner
- Professional liability insurance in excess of normal policy coverage, if required by owner

Project Administration Services

There are a number of activities associated with managing and administering an architectural project, including consulting with the owner, researching design criteria, attending meetings, communicating, and issuing progress reports. It is the architect's responsibility to coordinate his or her own services with those of the architect's consultants.

Additional project administration activities include:

- Furnishing a project schedule
- Considering the value of design, material, and equipment alternatives and their impact on program, budget, and aesthetics
- Explaining the design to the owner
- Submitting the design to the owner for evaluation
- Helping the owner file appropriate governmental documentation pertaining to the project

Architectural Programming

An architectural program, or brief, helps to define the scope, nature, scheduling, and cost of a project. It is the owner's responsibility to furnish a program setting forth project objectives, including space requirements and relationships, equipment, and site details. However, the architect will provide a preliminary evaluation of the information furnished by the owner pertaining to program, schedule, and budget, reviewing these for consistency with the overall intent of the project and determining if additional information is necessary to begin design.

For larger, more complex projects, the architect, or a consultant to either the owner or the architect, may be retained to provide programming services to assist the owner in articulating project requirements. Many public sector clients require such a programming phase. In these cases, the owner's further requirements should be established by obtaining information on:

- Design objectives and criteria
- Possible constraints
- Space requirements and relationships
- Future flexibility and expandability
- Special equipment or systems required
- Landscape or site requirements

Budget Evaluation

Once the requirements of the project are identified, the architect is also responsible for preparing a preliminary estimate of the project's costs. This may be calculated by:

- Area and volume method
- Unit use method
- In-place unit method
- Quantity and cost method

The method of cost calculation will depend upon the specific nature of the project.

If the estimate exceeds the owner's budget, the architect must make recommendations to the owner to adjust the scope, quality and/or budget of the project. Any evaluation of the owner's budget for the project, and preliminary estimates of the project cost does not imply or warrant that these will match the actual bids of contractors. The architect must also be allowed to make estimates based on contingency factors and alternates.

At this point, it may be useful to review potential methods of contracting for construction services. Although in some cases it may be too early to finalize decisions on the project, it is useful to consider the following matters together with their possible effects upon:

- Selection of the contractor (bidding or negotiation)
- Mode of project delivery (traditional, design-build, bridging, etc.)
- Form of payment to the contractor
- Use of bonus/penalty clauses
- Use of construction manager
- Rate of liquidated damages
- Employment of consultants
- Insurance, bonds
- Use of separate or single contract systems
- Additional architect's services

Surveys

According to AIA Document B141, Standard Form of Agreement between Owner and Architect (2.2.1.2), it is the owner's responsibility to provide all necessary descriptions of the site, and any further investigations which the architect considers appropriate. This may be done by employing the services of a land surveyor. However, the owner may seek assurance from the architect to undertake this work as part of the architect's services.

Specific data should be provided to enable the architect to make an adequate assessment of the site.

Site Surveys

Site surveys should describe the physical and legal characteristics of the site and should include (as applicable):

- Grades and lines of streets, alleys, pavements
- Adjoining property and structure
- Adjacent drainage
- Rights of way

- Restrictions, easements or encroachments
- Zoning
- Deed restrictions
- Boundaries and contours of the site
- Locations, dimensions and relevant data about existing buildings
- Trees and other improvements
- Utility services and lines (public and private) above and below grade

All information should be referenced to a project benchmark.

Note: Sketches, photos or video recordings of the site and surrounding area can prove useful for the architect as a ready-reference source back at the office, particularly if the site is some distance away.

Although provision of site information is generally the contractual responsibility of the owner, the architect should be assured of its accuracy, as reliance on outdated or inaccurate data might be attributed to architectural negligence.

Design Services

The architect's design services on big projects may include structural, mechanical and electrical engineering services, although more complex buildings will necessitate the engagement of consultants.

Design services are divided into these categories:

Schematic Design

The architect provides conceptual ideas for the project, illustrating the scale and relationships of the programmatic components. Materials may include:

- Site plan
- Preliminary building plans, sections, and elevations
- Study models
- Perspective sketches
- Electronic models
- Preliminary advice on major building systems and materials

Information that may be communicated at this stage may include:

- Location of proposed project on site
- Function and relationship of rooms and spaces (including their areas and heights)
- Primary elements (walls, floors, etc.)
- Overall appearance/character of the scheme

Design Development

Once the schematic design has been approved by the owner, along with an updated budget, the architect should start preparing more detailed illustrations and data related to the proposed design.

Any consultants who have been employed may give assistance at this stage, providing integrated input into the design process so that a final scheme can be prepared for the owner's approval, ready for the construction documents phase to begin. A further updated budget must also be submitted to the owner during this phase.

Submittals to the owner concerning the development of the design could include the following:

- Site plan
- Floor plans
- Elevations
- Sections
- Schedules and notes
- Calculations
- Preliminary draft of the project manual
- Outline specifications
- Other data (e.g., electrical, and mechanical systems)

The importance and significance of decisions made at this stage should be made clear to the owner, and written approval to proceed obtained before continuing to the next phase.

The Project Manual

During the design development phase, a project manual should be drafted. This will contain the bidding requirements and the contract documents, including the technical specifications, but excluding the drawings.

Construction Documents

When details of the project have been sufficiently determined and approved by the owner, the architect will undertake:

- preparation of detailed working drawings and specifications sufficient for construction purposes;
- assistance to the owner in securing bidding information, forms, contracts, and conditions (see page 67);
- any further changes in the projected construction cost;
- assistance to the owner in filing for any government approvals (see page 51).

Drawings

Some offices develop standardized practices in respect of working drawings and the construction documentation phase.

At the beginning of the documentation phase, estimate the number and type of drawings necessary to complement the specifications, and prepare a drawings schedule. This will facilitate office programming and improve the production sequence.

Draw only as much as is necessary. Time and money are often wasted in the duplication of material which is adequately covered elsewhere (e.g., in the specifications or schedules). Again, careful planning at the outset can help to prevent inefficiencies. Generally, information that relates to quantity and location should be in the drawings, while that pertaining to quality, method, and result should be in the specifications.

Use standard methods of cross-referencing throughout all projects so that employees become familiar with their use.

Although the number and mix of drawings will vary from project to project, the basic range of drawings is as follows:

Key drawings:
- site plan
- floor and roof plan
- elevations
- sections
- details
- schedules (e.g., doors, windows)

Structure and assembly drawing:
- foundation layout
- floor and roof layout
- sectional structural details
- relevant schedules (e.g., columns beams)

Mechanical drawing:
- mechanical layouts
- plumbing data and schedules
- heating, ventilating, and air conditioning layouts and data
- stacks

Electrical drawings:
- electrical layouts
- electrical details and schedules

Make sure each sheet contains:
- project title and location
- project number
- sheet number and title
- scale
- drafter's name
- checker's name
- date (of issuance and revision)
- north point (where relevant)
- space for stamp
- revision space
- the name and address of the practice
- space for owner's approval signature

Build up a collection of standard details or schedules that may be used in future schemes.

Specifications

Because of their important interrelationship, drawings and specifications should be developed together to avoid any duplication or omission of information. In accordance with the Uniform System for Construction Specifications, Data Filing and Cost Accounting (AIA Document K103), specifications are split into sixteen parts, or divisions.

The MASTERSPEC system may also be useful to the architect in the preparation of specifications: this is a computerized resource based upon the Uniform System and, at a time when it is becoming increasingly difficult to keep up to date with technical developments, represents a comprehensive and current professional aid.

Great care should be taken in the preparation of the specifications, which should generally either be handled by one of the principals or a specialist employed specifically for the purpose.

Specifications should be as concise and comprehensive as possible to prevent the necessity of too many addenda and/or the passing of alternates or unit prices (see page 68). Furthermore, the use of specific trade names should be avoided where possible, as this may unnecessarily restrict materials and product selection.

PRACTICE OVERVIEW

ARCHITECTURAL ETHICS

Almost a century of debate about architectural ethics reflects in large measure the evolution of the profession's view of itself as it has transformed from noble calling to competitive business. The result of those discussions codified in codes of conduct reflect the nature of professionalism in architectural practice.

What is a Profession?

Issues of ethical behavior inevitably raise the notion of professionalism and all it entails. While many would agree that architecture is certainly one of the major professions, along with other fields such as law and medicine, the definition of what that actually means is a little more elusive and bears some examination. Briefly, a profession is defined in *Webster's New Collegiate Dictionary* as "a calling requiring specialized skill and knowledge and often long and intensive academic preparation." More usefully, certain elements can usually be found in professions which may be absent in other career pursuits, including a restricted field of practice, a combination of academic and practical training, high degrees of professional autonomy and collective responsibility, a representative national body and enforceable codes of conduct.[1] It is the last category which is arguably pivotal to the operation of the professional body and which has caused problems in the last few years. Basically, codes of conduct, by providing specific guidelines for behavior, establish minimum standards for performance which determine the actions of each individual within the professional group in relation to their clients, their fellow professionals, and ultimately the public they serve. The codes are usually developed through consensus, and are particularly important to a profession such as architecture. Unlike medicine and law, architecture is not concerned with such intangibles as the maintenance of health and the preservation of freedom but with the provision of a consumer service. As such, it is more susceptible to the pressures of commercialism and the free market. Codes determining behavior in the architectural profession are established at various levels—by individual states, each of which maintains a licensing statute, by the American Institute of Architects and by the National Council of Architectural Registration Boards (NCARB). Although they cover some of the same areas, state requirements are specifically concerned with competence and the protection of public health, safety, and welfare, while the AIA and NCARB Codes delve further into professional behavior (and are, of course, only applicable to architects who are AIA and NCARB members respectively). Thus, the earlier AIA codes generally covered issues such as competence, plan stamping, and unauthorized practice which are usually found in state codes, but also provided, among other things, specific regulations dealing with competitiveness and client solicitation.

The Development of the AIA Codes of Conduct

While the AIA itself was formed in 1857, it did not develop its first ethics codes until 1909 when it approved, after considerable discussion, the "Circular of Advice Relative to Principles of Professional Practice and the Canons of Ethics" at its annual convention that year.[2] This document proclaimed that "Advertising tends to lower the dignity of the profession," although it was the ethics of competitions rather than advertising which held the profession's attention that year, self-promotion being considered more "an exhibition of bad taste rather than bad morals." Interest grew rapidly though, and the codes were revised regularly prior to World War II with the issue of advertising generating numerous and often violently conflicting discussions. In 1918, the canon preventing advertising was dropped after a particularly fierce debate, but was successfully reinstated in 1927 when the Principles and Canons of former years were forged into "Principles of Professional Practice."[3] By 1945, the rules stiffened considerably as high-minded proponents of advertising-free practice successfully persuaded their colleagues that restrictive codes were an essential component of a respectable profession. These codes stated unequivocally that "An Architect will not indulge in false publicity," and formed the basis for requirements which remained in place until 1978. By this time, they were very explicit in outlining what an architect could or could not do, particularly with regard to advertising and work solicitation. "Standards of Ethical Practice" (J330) established a number of canons that set forth obligations to the public, to the client and to the profession. They collectively codified a professional ethos that was intended to separate the architect from the more pragmatic attitudes and actions of the marketplace.

The End of the Mandatory Code

Unfortunately, the codes, as well as the codes of other professional bodies, began to raise issues of unlawful restraint of competition during the 1970s following a 1972 consent decree signed by the US Justice Department. This culminated in the Mardirosian case, which led to the demise of the code which had for so long helped to determine architectural behavior. Briefly, matters came to a head when a Washington architect was suspended for one year from the AIA in 1977 for alleged violation of the supplanting clause. He challenged the decision on the grounds that the AIA was imposing a restraint of trade which violated the Sherman Anti-Trust Act. When a federal District Court judge ruled in his favor, the plaintiff was reinstated into the Institute and received a $700,000 out-of-court settlement. The AIA promptly withdrew the mandatory codes before any further challenges were brought and replaced them with a set of purely voluntary guidelines, "Ethical Principles" (63400), which were published in 1980. This move effectively transferred the conscience of the profession from the collective level to the individual level in matters concerning advertising and job competition and led to some interesting developments. Advertising was always the most common infraction of the old rules,[4] so the lure of unrestricted opportunities led some firms to dabble in full page ads and promotional exercises which drew reactions from their colleagues ranging from mild bemusement to high dudgeon. Additionally, at least one regional director was forced to deal with a practice in his state which was systematically contacting the clients of other architects and offering to work for lower fees. Similar unsettling incidents were rumored, although no official action could be taken as the profession was no longer allowed to regulate such behavior.

The "Code of Ethics and Professional Conduct"

The broad and rather vague contents of the "Ethical Principles" were felt by many to be too insubstantial to give members adequate direction, so, after three years of careful preparation, the AIA published the current "Code of Ethics and Professional Conduct" which was intended to provide more specific guidelines without violating federal law. The document is divided into *Canons*—broad principles of conduct, *Ethical Standards*—specific goals towards which members should aspire, and mandatory *Rules of Conduct*, transgression of which can lead to disciplinary action by the AIA's National

Judicial Council. This body interprets and enforces the code, although it does not become involved in fee disputes or cases of client dissatisfaction with a member. It does, however, issue advisory opinions which are nationally distributed to AIA Chapters to provide some guidance on ethical issues.[5] However, all units of the organization are acutely aware of the dangers of establishing rules which transgress any part of anti-trust legislation. While establishing some much needed structure and guidance, the code by necessity avoids any restrictions on practice behavior which, it has been argued in the past, goes to the very heart of professionalism—the means by which architects deal with their clients and their peers, and how they present themselves to society.

Is There an Ethics Problem in the Profession?

While the code has raised some issues with regard to its implications for competence and how that relates to liability,[6] it still ensures that the choice of architects' attitudes towards advertising and job selection remain solely a matter of personal choice. Given the pressures of commercialism, have members stepped over the boundaries of behavior established in former codes and threatened the notion of professionalism in architecture?

Some members of the profession evidently believe so in the wake of code change. The results of a poll conducted among readers of *Progressive Architecture*[7] indicated a general concern about ethical abuses. However, no clear-cut agreement on correct ethical behavior emerged, except in obvious areas like public safety, the unauthorized stamping of drawings, the embellishment of credentials, and the padding of bills—all of which are covered under the new code and suggest problems of enforcement, not omission. Interestingly, concerns expressed by older architects diverged from those of younger members. The latter perceived a greater amount of unethical behavior and focused, not surprisingly, on low pay for recent graduates and false promises of advancement as areas of greatest concern. Older practitioners for their part felt that moonlighting was a problem, although the notion of accepting gifts from contractors—forbidden under the old rules—seemed to be relatively acceptable to both age groups.

The profession was split on the issues of advertising and solicitation of work from a client who has already agreed to work with another architect. Younger practitioners distinguished themselves by, on the one hand, disapproving of advertising more than their older counterparts, but being more receptive to the idea of competition in client solicitations. In both instances, though, respondents did not collectively express strong objection to either area as a major source of ethical breach.

Perhaps this indicates a softening of professional attitudes towards practices which are considered commonplace within the rest of the commercial world, suggesting that the restrictions formerly maintained on advertising and client solicitation were outdated and inappropriate. Certainly, such arguments have been made by some members since codes first appeared.

Alternatively, perhaps the more restrictive codes were not really necessary. While the flurry of advertising which came out in the wake of the Mardirosian case caused a stir, it was a fairly short-lived phenomenon, and most examples seen these days are mild in content and limited in distribution. Many architects became aware that "big bang" advertising was not the most cost-effective way to secure commissions. Similarly, there appear to have been limited problems with supplanting in the last few years. While some complaints have been made to the AIA, the considerable practical problems associated with taking over another architect's work—liability, copyright, the need for hold-harmless clauses, etc.—seem to have largely obviated the need for additional pressure.

Summary

As codes of conduct developed in the architectural profession, the issues of advertising and solicitation, once the fulcrum of great debate and contention, faded from interest as professional attitudes have evolved. Despite fundamental changes in the codes precipitated by actions of the judiciary, the removal of restrictions addressing the two areas do not appear to have created any major threats to professional practice. However, this is not to say that problems do not exist or that the current code is perfect. Approximately 78 percent of the respondents to the *Progressive Architecture* poll believe that, despite reinforcement, it is still insufficient to maintain necessary standards of ethical behavior, notwithstanding the lack of any substantial evidence of major problems in the profession. Perhaps this indicates a desire among architects for greater restrictions and clear-cut boundaries demarcating professional behavior. Perhaps it is a fear of too much unrestrained competition or perhaps simply a desire to see high personal standards implemented throughout the profession. Whatever the reasons for concern, it is unlikely that debate on the issues has ended. Regardless of legal and economic pressures, the appropriate ethical behavior of architects has been a topic of lively discussion for nearly one hundred years and will inevitably surface again at some future AIA Convention.

References

1. Ware, P., "The Sociology of the Professions," Chapter Two.
2. Osman, M., "To Advertise or Not to Advertise? Footnotes from History," *AIA Journal*, December 1978, 55–7.
3. Coxe, W., *Marketing Architectural and Engineering Services* Van Nostrand Reinhold, 1971, 9–35. The book contains an excellent account of codes relating to advertising since the beginning of the century.
4. Coxe, W. *op. cit.*
5. The National Judicial Council has issued two Advisory Opinions: (1) Misleading Prospective Client—Uncompensated Design Services, 30 June 1987; (2) Conflict of Interest—Referral Fees, 30 June 1987.
6. Coplan, N., "Law: AIA Code of Ethics," *Progressive Architecture*, March 1987, 61–5.
7. *Progressive Architecture*, February 1988, 15–9.

Question & Answer

Is getting a license really that important to an architect, especially as I have a Masters degree and several years' experience?

Architecture is a restricted area of practice in all states, although the definition of what constitutes "architecture," "the practice of architecture," or the title "architect" may vary. For example, any project under 50,000 cubic feet that does not affect public health and safety does not constitute architecture in many states, which excludes the majority of housing work.

Anyone wishing to use the title "architect" must be registered in each state where practice is undertaken and abide by the rules established in each administrative state's code (see page 31).

An important aspect of licensure is the ability to approve design and construction drawings by the imprimatur of a stamp. All projects are required to be under the supervision and control of the architect, who will bear the responsibility for any failure attributable to design failure.

Anyone wanting to practice as an architect must meet state requirements which frequently require:

- an accredited degree;
- a period of internship;
- successful completion of a rigorous examination.

Some, but not all, states offer reciprocity to enable practice outside of the architect's state of registration. The National Council of Architectural Registration Boards provides a nationally recognized level of qualification through its requirements that facilitates practice beyond the confines of the home state.

Law and the design phase

PROPERTY LAW

Certain legal rights, obligations, and restraints concerning land should be considered in the design process as they may affect:

- the choice of site for a particular development;
- the character of the development of the site that has been selected;
- the methods and procedures to be adopted in any proposed development.

Land law varies from state to state in its details and specific applications, but some general observations on a few important aspects of land law can be made.

Ownership

Ownership of land is expressed by *title*, which can be transferred from one party to another. Prior to purchase, a prospective owner normally has an investigation carried out into the background of the property:

- to ensure that the prospective seller actually possesses a transferable title to the land;
- to check whether any encumbrances are attached to the title which might affect the future enjoyment of the land;
- to discover whether any governmental statutes or regulations exist which restrict development or usage of the land (see page 51).

Transfer of ownership is accomplished by *deed*, of which there are three basic types:

1. General warranty
2. Special warranty
3. Quitclaim

General Warranty

By this kind of transfer, the transferor remains personally responsible for the title indefinitely. Although this is the most secure form of deed from the buyer's point of view, it is rarely granted.

Special Warranty

This guarantees that the land has not been encumbered during the current ownership, but gives no assurances in respect of the title prior to that period. This is the most common form of transfer deed, and its character requires that the title be carefully investigated.

Quitclaim

If a quitclaim is used, the owner promises nothing except that he or she will not contest the new ownership. This form of transfer is inadvisable in most circumstances.

Absolute and Acquired Rights in Land

Certain rights accrue to the owner of the land which require no legal formality beyond the transfer deed (e.g., lateral support). Other rights can be of great importance, but must be formally acquired. These rights are often created by easements or covenants.

Easements and Covenants

These are legally enforceable, and attach certain conditions to specific land.

Easements are legal rights enjoyed by one party over the property of another. They are usually described in a deed or in a license, but in some circumstances they can be implied by usage over a long period (e.g., five to thirty years' continual use has been considered sufficient to imply an easement). Easements are frequently sought with regard to:

- Access
- Light
- View

Restrictive covenants restrain an owner from undertaking certain actions in relation to his or her land. They are usually established by a previous owner (e.g., the developer, if the land forms part of a general development) and they are often introduced in an effort to protect the character of a neighborhood, or to maintain property values. Restrictive covenants may be used:

- To prevent fence building
- To assure minimum levels of aesthetic or architectural appearance
- To prevent major alteration or change to existing buildings
- To prevent tree-lopping

In certain circumstances, easements can be sought by a new owner in connection with the property of a neighbor (e.g., in order to dig out foundations close to the boundary). A license for this purpose should be requested and may involve monetary consideration.

A prospective purchaser of land should ensure that a thorough check is carried out by a suitably qualified professional, to familiarize the buyer with any easements or covenants which may exist and which may affect the land's usage. This investigation should also include searches for other encumbrances (e.g., unpaid mortgages, outstanding liens, etc).

Other legal provisions exist which can affect the relationship between neighbors, and the liability for persons entering upon the land. These include:

1. Spite fences
2. Tree ordinances
3. Nuisance
4. Occupier's liability

Spite Fences

In some states, if a landowner maliciously erects a high fence which interferes with a neighbor's land (e.g., by causing excessive shading or view blocking), the courts can order the fence to be removed and allow the offended neighbor to claim damages. Other states refuse to interfere in the case of a spite fence on the basis that a landowner should have free use of the land owned. However, there appears to be a trend toward some control of spite fences.

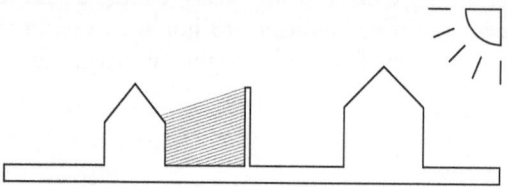

Figure 4.1

Tree Ordinances

In some localities, where a view is blocked by excessive foliage from a neighbor's tree, a reasonable request can be made to remove the obstruction, with costs to be shared between the two parties. Tree Commissions are sometimes set up to decide these matters in case of dispute.

Nuisance

Ownership of property generally entitles the owner to enjoy the land without interference by neighbors. Sometimes the activities of one party affect the enjoyment of the other to the extent that legal action might be taken to prevent further disturbance. An action for nuisance might be brought in respect of:

- loud noises;
- antisocial activities (e.g., excessive vibration caused by pile-driving);
- pungent odors or smoke;
- unsightly appearance of neighboring property.

It is important to remember that nuisance is classified as a tort, and in each case the court is concerned to discover whether a reasonable person would find the act complained of to be disturbing. The court will often also consider the conduct of the plaintiff and the extent and duration of the alleged nuisance in reaching its decision. Another consideration is the benefit to society at large (e.g., industrial disturbances) which will be weighed against the disturbance caused to the plaintiff.

Occupier's Liability

Occupiers have a duty of care to all persons lawfully on their premise, and the duty varies according to the classification of the visitor:

Invitees

These are owed the highest duty of care by the occupier, who is responsible for those hazards known to exist, and those which could reasonably have been revealed. The category of invitees does not include social guests.

Licensees

A licensee is generally a person who comes onto the premise for personal reasons rather than for the purposes of the occupier, but with the occupier's consent (e.g., sales representatives). Social guests fall into this category. The occupier is obligated not to subject licensees to unreasonable risks, but this duty is reduced if licensees are in any way partially responsible for the injuries they sustain.

Passers-by

Boundaries should be clearly demarcated, and activities on the property conducted so as to show reasonable care in avoiding injury to passers-by.

Trespassers

Trespass may be defined as the unauthorized transgression of another person's land, including the airspace above and the ground below. Trespassing is classified as a tort (see page 6).

Even individuals who enter premises as trespassers are owed some duty of care, although this is reduced, particularly if the trespasser is intent on malicious damage. However, attempts to physically abuse the intruder should be limited to personal protection: courts have in the past held that protection of property alone does not justify extreme physical assault upon a trespasser, and in some cases high damages have been awarded against the occupier.

The duty to child trespassers is generally higher than to adults because children may be less aware of property boundaries and inherent dangers than adults. In particular, building sites should be adequately secured and posted.

The above classifications are highly technical and it is difficult for ordinary persons to be sure of the status of all people who enter on their land. For this reason, occupiers should exercise great care in keeping their premises reasonably safe.

GOVERNMENTAL RESTRAINTS

In the early stages of each project, attention should be paid to the various legal restrictions which might affect the scheme. Some restrictions apply to all projects, whereas others are applicable only to projects of a certain type, or in a particular location.

General Provisions

Although these vary according to state and locality, most building operations will require:

- a zoning permit (see page 52);
- a building permit (see page 56);
- services connection (e.g., gas, electricity, water, sewage, telephone, etc.).

Special Provisions

In addition to the general requirements, some projects will require additional attention:

- OSHA
- HUD

OSHA

The Occupational Safety and Health Administration was set up by an Act of 1970 which makes it illegal to work in an unsafe place. OSHA has the authority to enforce safety standards, and to impose high financial penalties for violation of safety regulations. It enforces its standards by carrying out inspections of workplaces where accidents have occurred, or from which complaints have been made by employees. Spot checks are also made by OSHA officials to promote widespread adherence to the regulations.

OSHA Checklist Architects are particularly concerned with section 1910 of the Occupational Safety and Health Act. The following points help to avoid problems with conformance to the OSHA regulations:

1. Obtain clear instructions of intended use from the client so that OSHA provisions can be considered in the early stages of design.
2. Ensure that the client is aware that compliance with OSHA can unavoidably increase the cost of the project.
3. Note that, in the case of a conflict between OSHA provisions and local building codes, the more stringent regulations prevail.

Other governmental departments may have additional requirements. For example, the Department of Housing and Urban Development was created in 1965 to help alleviate problems in urban areas by the promotion of major federal aid programs coupled with financial aid and technical assistance. If HUD is providing an input to a particular project, design guidelines may be laid down with which the architect is expected to comply.

Specific Types of Project

Particular projects are sometimes affected by specific legal constraints and need approval and/or inspection by individual state authorities which may impose their own standards (e.g., hospitals, schools, factories, etc.).

Location

Some restrictions affect all proposed projects within a specified area, particularly districts designated as historic preservation zones. In these, locally elected commissions develop and enforce rules and standards for future development. In some cases, individual buildings are singled out as

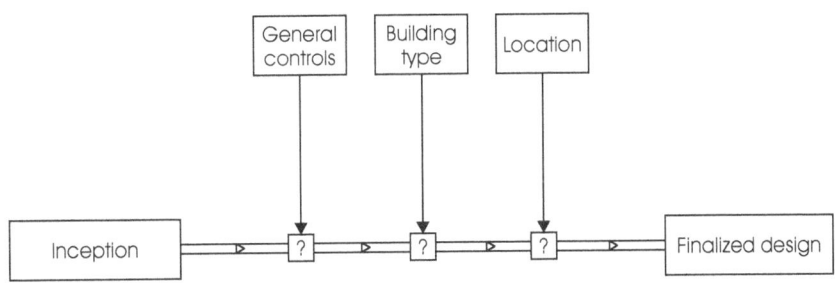

Figure 4.2

historic landmarks, either at a federal, state or local level, and this effectively prevents their demolition or alteration unless an appeal is successfully made according to the designated procedures. Architects working in older districts or on the alteration or extension of older buildings should check first to discover whether the preceding restrictions have been imposed, and consider how extra requirements will affect the design and progress of the project.

In assessing each project at the outset, the architect should ensure that the owner is aware of the scope of the architect's services in respect to gaining approvals so that there is no misunderstanding should there be a need for additional payment (e.g., where a zoning appeal is necessary). It is most inadvisable for an architect to assure a client that the necessary approvals will be granted without difficulty. Attention to these matters at the early stages of the project will help to prevent any later decline in the architect/owner relationship.

ZONING

Definitions

- *Zone:* to mark off into zones; specific, to divide (a city, etc.) into areas determined by specific restrictions on types of construction, as into residential and business areas. (*Webster's New World Dictionary*).
- *Zoning Permit:* a permit issued by appropriate governmental authority authorizing land to be used for a specific purpose (*AIA Glossary of Construction Industry Terms*).

General

Zoning activity is the responsibility of individual states which pass zoning legislation as part of their police powers to protect the community. United States' zoning originated in the enabling acts of the 1920s which placed power to create and administer public land use regulations in the hands of local authorities. The most common type of zoning became known as Euclidean zoning; it consisted of establishing specific districts or zones for particular uses, e.g., commercial, manufacturing, residential, etc. These zones were then broken down into smaller units, e.g., light and heavy industry.

Scope

In addition to restricting use, the zoning regulations grew to cover matters such as:

- Density
- Light

- Air
- Space
- Height of buildings
- Bulk of buildings
- Plot sizes
- Aesthetic considerations

Model Land Development Code

Although the Euclidean model was, and is, in common usage, cumulative zoning (i.e., allowing carefully regulated, multi-use districts) has gradually developed since the 1920s. In 1975, the Model Land Development Code was approved by the American Law Institute. The MLDC is only a recommended code, but some authorities have adopted some of its recommendations which include:

- substantial responsibility for administering the development scheme should lie with local authorities;
- state authorities should provide some input to avoid state problems resulting from purely local administration;
- less rigid approaches to zoning;
- more stress on the environmental and aesthetic considerations of zoning.

Procedures

In many communities, the building department deals with zoning requirements, although where size and complexity of a community warrant a separate administration, zoning officers are sometimes appointed.

Applications for zoning and building permits are made simultaneously. If the applicant's request for a zoning permit is rejected, appeal procedures are generally available (see page 53). However, an appeal may not be the only alternative if a proposed project fails to match zoning requirements: procedures allowing greater zoning flexibility are available in many localities. These include:

1. Variances
2. Special use permits
3. Conditional permits
4. Rezoning

Variances

Variances may be granted to enable land to be used for a different purpose than the category stated in the zoning ordinance. Specific requirements vary, but generally the applicant must show:

a. that exceptional circumstances exist;
b. that strict application of the zoning ordinance would result in hardship;

c. that the granting of the variance would not be detrimental to the public at large, or to those owning neighboring property.

Other restrictions often apply to the granting of variances.

Special Use Permits

Many localities make provision for the issuance of special use permits in given circumstances, which may be expressed in specific or general terms by the zoning ordinance.

Conditional Permits

In some cases, permission may be granted by the zoning authority contrary to the ordinance, provided that the applicant agrees to fulfill certain conditional requirements (e.g., noise control, provision of fences, etc.)

Rezoning

When an owner cannot match either the zoning requirements, or the conditions for a variance or a special use permit, application may be made to have the area in question rezoned. This is a difficult procedure especially if, as is often the case, neighboring landowners would suffer hardship as a result.

Common Features

Zoning is a complex and detailed field which can vary considerably from place to place. Care should be taken to gain an understanding of the zoning law which applies to the locality of a proposed project. However, there are several features common to many local zoning policies including:

1. Nonconforming uses
2. Floating and bonus zones
3. Environmental impact statements
4. Green belt/open space zoning
5. Conservation of historic buildings and landmarks

Nonconforming Uses

Where an existing use failed to conform with new zoning segregation policy, the tendency was to eliminate that use. This approach caused considerable problems, and now several areas allow nonconforming uses to continue if they were lawful and existing prior to the new zone being established. Conditions for nonconforming uses vary from one locality to another.

Floating and Bonus Zones

Floating zones are sometimes located within specified zones to provide a measure of flexibility in future development. Bonus zones allow possible dispensation from the requirements of the zoning ordinance, provided that certain extras or "bonuses" are built into the project for the benefit of the community. For example, certain buildings in New York have been allowed to violate aspects of the ordinance on condition they provide public plazas or shopping arcades.

Environmental Impact Statements

These have been developed in some areas as a means of protecting and improving the environment by requiring detailed accounts of probable environmental consequences of certain zoning decisions. The statements are concerned with issues such as:

- Pollution
- Natural resources
- Coastlines and scenic features

Green Belt/Open Space Zoning and Smart Growth

This type of zoning is gaining support in the United States and it provides for the maintenance of open spaces, free from development and restricted to specific activities, e.g., recreation.

In recent years, states have enacted Smart Growth legislation, which attempts to control suburban sprawl, enhance public transportation and encourage sustainable growth.

Conservation of Historic Buildings and Landmarks

Increasingly, federal, state or local government authorities are taking steps to preserve historic districts, or individual historic buildings. Where this type of land use control exists, there are often provisions to alleviate the possible financial burden on the owner.

Appeals

A Board of Zoning Appeals is usually established in each locality and given the power to modify, reverse, or uphold zoning decisions. Appeals Boards are also usually empowered to grant variances and special or conditional permits. The duties and powers of the local Appeals Boards and the procedure which they must follow vary from place to place, but are generally defined in the relevant ordinances.

Appeal Procedure

Generally a Notice of Appeal must be made on the appropriate form which may require information and enclosure such as:

- Name and address of appellant
- Identification of property
- Name and address of agent (if any)
- Affidavit of appellant or agent
- Date of decision appealed
- Proposed use of property
- Present use of property
- Zoning classification
- Estimated cost of construction
- Copy of the decision against which the appeal is made
- Statement of grounds of appeal
- Certified plan survey
- Plans and drawings of the scheme
- Proof of ownership
- The requisite filing fee

Other information and enclosure may be required, and some documents may have to be submitted with a specified number of copies. Generally, all zoning appeals must be made within the time limit stated in the relevant ordinance.

In the event that the appellant is unsuccessful in the appeal, there may be a further application to the regular courts in certain limited circumstances (e.g., if the Board of Zoning Appeals acted beyond the scope of its authority).

BUILDING CONTROL

Most construction projects have to conform to the requirements of the local building inspectorate,

which are expressed in the form of a building code. Building codes vary in their scope and coverage, but tend to concern the following areas:

- Health
- Safety
- Welfare

Typical codes include sanitary provisions, fire protection, structural requirements, etc. To ensure the code provisions are complied with, building permits are required for all building work (with a few exceptions: see page 56).

Buildings are divided into use or occupancy groups according to their proposed purpose. Each category has a separate set of requirements which must be matched in addition to the general provisions which apply to all building work. The occupancy groups may include:

- Assembly
- Business
- Factory and industrial
- Institutional
- High hazard
- Mercantile
- Residential
- Storage

Although a number of states have enacted statewide provisions for certain types of building, building control in the United States is predominantly carried out under the authority delegated by each state to the individual locality. Although

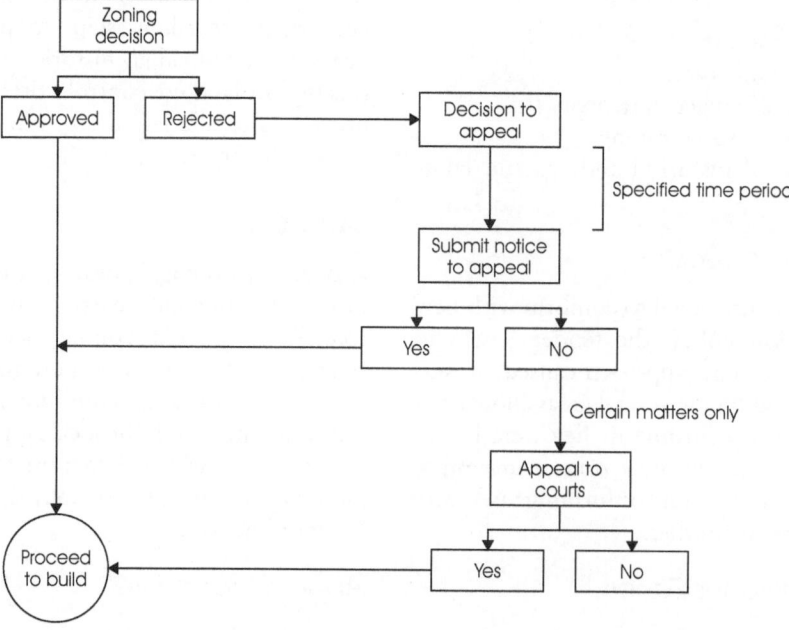

Figure 4.3

the level of uniformity has been raised by the development of the International Building Code, this means that each local authority can develop and administer its own individual code, which generally consists of several sections, dealing with:

- Building
- Electrical
- Plumbing
- Heating, ventilation, and refrigeration
- Housing
- Fire

In addition, most localities have separate Fire Prevention Codes which are administered by the local Fire Department.

Types of Code

Building code requirements can be expressed in different ways as:

1. Specific regulations
2. Functional requirements
3. Performance standards

Specific Regulations

There are basic statements of a direct nature, giving a fixed and clear standard providing a limited range of solutions and little flexibility of choice.

Functional Requirements

These give complete freedom to provide a solution by making very generalized requirements, without indicating how the desired level of performance might be achieved (e.g., buildings should be designed and constructed so that if a fire breaks out, everybody can evacuate the building and immediate area in safety). Despite their flexibility, functional requirements are sometimes criticized because of the lack of direction in their demands.

Performance Standards

These provide an intermediate alternative to the other forms of regulation by providing a measurable and precise account of the performance that is required, but leaving it to the designer to decide.

The Model Codes

In an effort to assist local authorities in the development of building codes, several model codes were developed and published by groups concerned with the building control process, including:

- The Uniform Building Code, published by the International Conference of Building Officials (ICBO)

- The National Building Code, published by the Building Officials' Conference of America (BOCA)
- The Southern Standard Building Code, published by the Southern Building Code Congress International

Recently, the three major building codes have been consolidated into a single model code known as the International Building Code, published in 2000.

The model codes are widely used in the regulatory process but, owing to their voluntary status, each authority may amend, alter, or ignore the models as they wish. Out-of-date provisions have also been identified as a problem. Architects working over a wide geographic area should take care in checking the regulatory requirements specific to each project during the early design stages and, in the event of uncertainty, contact the relevant building department.

Standards

The building codes are a tabulation of accepted standards in building practice, and are developed by reference to organizations which test and give official approval to new building materials and techniques. The opinions of each organization are not binding upon local authorities, but the more respected of these tend to be widely accepted, including:

- The American Society for Testing Materials (ASTM)
- The National Fire Protection Association (NFPA)
- The American National Standards Institute (ANSI)
- United States Department of Commerce (USDC)

Accessibility Guidelines

Various forms of federal civil rights legislation, including the Americans with Disabilities Act (ADA) and the 1988 Federal Fair Housing Amendments Act, ensure accessibility to a wide variety of settings. Individual states may also enact more stringent requirements.

ADA

The Americans with Disabilities Act Accessibility Guidelines (ADAAG) 1990 establish requirements for accessibility for public accommodations, including governmental offices, private businesses, and public transportation facilities. The guidelines are developed by the US Architectural and Transportation Barrier Compliance Board (ATBCB) and are enforced by the US Department of Justice (DOJ). Some state and local building codes have been certified by DOJ as equivalent. However, ADAAG are sufficiently ambiguous in places to warrant diligence in their interpretation. DOJ provides technical assistance through various materials available through its website and a telephone hotline: 1-(800)-514-0301.

1988 Federal Fair Housing Amendments Act

This legislation applies to certain types of privately-owned multifamily housing. It ensures accessibility of public accommodation spaces and the adaptability of dwelling units. Specific requirements are enumerated in the Fair Housing Accessibility Guidelines adopted by the US Department of Housing and Urban Development (HUD). States may also enact their own adaptability legislation.

BUILDING PERMIT APPLICATION

Each locality administers its own building regulations and specifies the procedures which applicants must follow to obtain a permit. Outlined here are some common aspects of the procedures followed in many localities, but architects and developers should take care to discover the application procedures relevant to the locality of the project. Any questions should be directed to the relevant building department. Architects require written authority of the intent to make a permit application as the landowner's agent.

Need for Permit

Most building work requires a permit including:

- Construction
- Demolition
- Occupancy
- Heating and cooling installation
- Moving of buildings
- Alterations

Exemptions are limited to minor alterations and ordinary repairs where:

1. no work of a structural nature is proposed, and
2. health and safety are not affected.

Applications for permits usually consist of the following:

- The completed form of application which includes a description of the work, location, occupancy use, and other information required by the individual building department

- A plot survey showing proposed and existing buildings, distances to lot lines, and accurate boundary line information, etc.
- Building plans including drawings and specifications of sufficient size and detail to show the character and nature of the work (foundation plans, floor and roof plans showing exits, etc.)
- Additional details, e.g., structural, mechanical, electrical drawings, computations, stress diagrams etc., structural calculations and provisions for fire resistance
- The required fees

The building inspectorate will consider the application and either:

- issue a building permit;
- reject the application in writing, specifying reasons for the rejection.

If a permit is issued, work must generally begin within a fixed period and be completed within a specified time.

A notice signed by the building inspector confirming the issuance of the permit must generally be displayed on the construction site.

The building inspector has the right to enter the site at any reasonable time to check that the work is in compliance with the code. Required inspections are likely to be undertaken at important stages in the construction process, and due notice must be given to the inspector before such work begins. These stages may include:

- Foundations (trench, reinforcement, weatherproofing, etc.)
- Concrete slabs and framing
- Roofing
- Electrical work
- Gas piping and fixtures
- Plumbing and sewer connections
- Heating, ventilation, and refrigeration
- Plastering (interior and exterior)

The cost of tests which are ordered by the building inspector in connection with these duties must be borne by the owner.

Certificate of Occupancy

When the work is completed, a *certificate of occupancy* must usually be obtained before the building can be used. This certificate confirms that all the building regulations have been complied with, and must be available for inspection of the premises at all times.

If a part or portion of the work is ready for occupation before the completion of the project as a whole, a temporary certification may be issued for the part of the work concerned.

Variations

Where there are practical difficulties in carrying out the requirements of the building code, the inspectorate may, upon written request, vary or modify it as long as the spirit of the law is upheld, and health and safety provisions are not affected.

Violations

If at any time during the construction process the inspector feels that the work is not sufficient to satisfy the building code, the contractor will be required to amend the work before further certification. In certain instances, the building permit can be revoked, and a Stop Work Order may be issued if the inspector considers the violation to be sufficiently serious.

Appeals

If an application for a building permit or a variation is rejected, there is generally provision for appeal to a Board of Appeals which is empowered to uphold, reverse, or modify the inspector's decision.

If the appellant is not content with the Board's ruling, a court action might be considered if there are sufficient grounds. Advice of legal counsel should be sought before such action is taken.

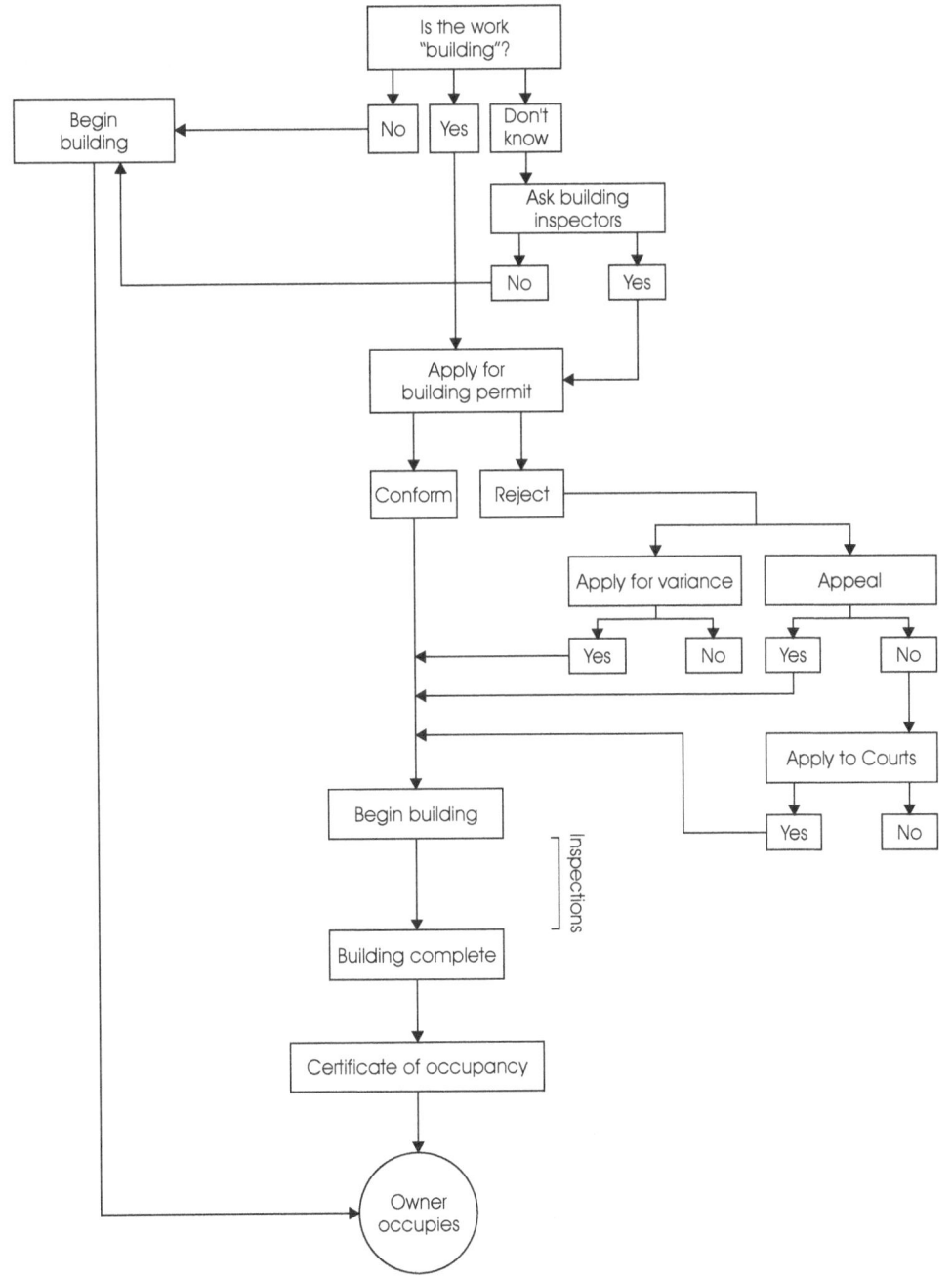

Figure 4.4

PRACTICE OVERVIEW

THE CASE FOR COPYRIGHT

The 1980s sparked some interesting cases involving architectural copyright, peppered with well-publicized disputes involving such personalities as Steven Holl, Donald Trump, and Arquitectonica. These cases revolved around the argument that architects provide a service, not a product, and therefore ownership of the ideas embodied in the end result—the buildings—could not pass to the owner without specific agreement.

Interesting as the issue was, most of the cases and, ultimately, interest fizzled out until the latest attempt to create legal safeguard. The US joined the Berne Convention in 1989 and, to align with its international provisions (which hold copyright as a natural rather than a statutory right), Congress enacted the Architectural Works Copyright Protection Act of 1990. The new act replaces legislation that primarily protected the drawings (rather than the embodied ideas), and has now been in place long enough for us to assess its effectiveness.

Small Scale

The 1990 act provides valuable protection for architects in a specific condition. It prevents their designs as well as their drawings from being reused without their permission or compensation. However, it has raised some interesting questions as to the definition of "architectural works" (for example, churches and gazebos are included, but parking garages, grain silos or even freestanding walls may not be), what actually merits copyright protection and, most interestingly, what constitutes real originality.

The act states specifically that copyrightable matter must be "an original work of authorship," although quality, aesthetic merit, ingenuity, and uniqueness are not necessarily factors. What is important is that the work must contain a "certain minimum amount of original creative expression," and that copyright registration cannot be based on standard designs such as common architectural molding or features, nor upon design elements that are functionally required. This creates a wealth of opportunity for dispute, particularly in smaller projects with few design variables such as houses, where permutations of bathrooms, kitchens, structural walls, windows, etc.—all arguably functional requirements—are relatively limited. However, it is in the house building industry where the issue of originality seems to be most intently debated. This is ironic, as housing is not a field traditionally dominated by architects (the American Institute of Architects once estimated that as little as 1 percent of American single family houses were designed by architects), nor one celebrated for widespread design originality, but it is the housing realm where issues of originality may ultimately be decided.

Three recent disputes focus on the same scenario: A home-building company applies for and receives copyright protection for their model home styles—the "Lakeside Colonial," the "Traditional Saltbox," etc.—and then sues another home-building company that subsequently builds something strikingly similar. The cases, none of which have yet found resolution in the courts, suggest a major shift in home-building habits and create some potentially interesting implications for architects in particular and the design industry as a whole. First, the notion of jealously protecting the design integrity of, say, the "Lakeside

Colonial" tends to fly in the face of traditional house-building habits of the past century. House plans and styles have been published freely in newspapers, journals, and specialty magazines beginning in the 1920s—even Frank Lloyd Wright once published some model houses for general consumption—with the intention of giving owners alternatives to use when discussing a new house with a builder. (And, of course, the discussion of a particular model or housing type, with or without modifications, is just as likely to involve the brochures of numerous home builders collected by the prospective owner.)

Second, the kinds of works submitted for and receiving copyright protection scarcely fall into the category of cutting-edge design, limited as they are in scale, budget, and, in many cases, architectural expertise. Furthermore, despite the best intentions of the act to prevent flagrant, wholesale copying of existing designs and drawings, how can protection on the grounds of originality be given to a colonial or a saltbox? Aren't they by definition redolent of generic styles that have long been in existence?

Large Scale

Precedents now being determined in the home-building end of the copyright spectrum may also affect the architectural profession beyond the singular building to the physical environment as a whole. While copyright protection shields the rights of individuals on a building-by-building basis, it cannot deal with the notion of multiple buildings, the issue of precedent, or the need to create physically coherent communities.

Sometimes, being a good neighbor—blending in with the existing context of buildings—is an appropriate response and one certainly taught as a relevant strategy in architecture schools. If copyright law vigorously protects the design uniqueness of each building, then each new building, it might be argued, has to be designed to consciously be different from every other—not a recipe for a coherent built environment. Illustrative cases include the Trump Tower, where changes to the nearby building were legally mandated to prevent its appearance from being too similar to the original "statement;" this, despite urbanistic argument that the towers together could create a powerful and coherent gateway to the street and the neighborhood.

This would not be the first time case law—the law as defined by the courts—created situations never conceived by drafters of the original legislation. If case law becomes problematic, there is of course the recourse of new legislation, although this is a slow, cumbersome, and equally unpredictable course. The best strategy for architects and planners for the time being is to stay informed, stay within the architects' standard of care, and continue to strive for the originality of creation that drew architectural works copyright protection in the first place.

Law and the design phase

Law and Practice for Architects • 59

Question & Answer

I love to experiment with new ideas and materials, and it often saves my clients money if I use a less expensive product that has just come on the market. Am I incurring any unnecessary risks in veering from the tried-and-true traditional methods and materials?

Selection of building materials and components is the responsibility of the architect and, if they subsequently fail, may result in liability. While this may be less of a worry with traditional, tried-and-tested solutions, the use of innovative, new materials may expose the architect to claims of negligent specification should they fail. The failure of high alumina cement, for example, which, despite rigorous preliminary testing, proved to deteriorate drastically after a number of years, caused many problems to the professionals who relied on its early promise.

While new products and materials can provide many benefits—savings to the owner, exciting new design possibilities, etc.—architects could minimize exposure by incorporating the following guidelines in their selection of relatively unproven building components:

- Insist upon detailed manufacturer's information and test results.
- Require a list of previous users and contact them, particularly those involved in similar projects.
- Require the approval of nationally recognized standards institutes.
- If still in doubt, require further independent tests.
- Inform the manufacturer of the proposed use of the product and ask for written comments on its suitability.
- Request warranties from manufacturers.
- Retain all written representations, communications and warranties well beyond the completion of the project.
- If the installation of the material is unusual or specifically defined by the manufacturer, request that their representative be present during installation.

Contract formation

CONTRACT LAW

A contract has been defined as:

> A legally binding agreement between two or more parties, by which rights are acquired by one or more to acts or forebearances on the part of the other or others.
>
> (Sir William Anson)

Formation

Contracts may be formed in a number of ways:

a. *Orally:* A contractual relationship may be formed between parties in some instances where no written agreement exists but a verbal contract was made.

b. *By conduct:* The actions of parties may be such as to imply a contractual relationship between them.

c. *In writing:* Certain kinds of contracts should be formed in writing if they are to be enforceable (e.g., in some states, real estate brokerage service contracts must be in writing).

d. *Under seal:* Contracts made in this form traditionally did not require consideration to be enforceable in the courts (see below). However, the law relating to the status of sealed contracts now differs greatly from state to state.

e. *Evidenced in writing:* Some contracts must be evidenced in writing if they are to be enforceable (e.g., contracts which, according to their terms, cannot be performed within one year).

Status

Contracts may be:

- Valid
- Void—without any legal effect
- Voidable—i.e., valid until one of the parties repudiates
- Unenforceable—in the courts

Elements of a Valid Contract

There are a number of basic elements which are necessary for the creation of a legally binding and enforceable contract. These are:

1. Offer and acceptance: An offer by one party must be clearly made, and the offer must be unconditionally accepted by the other party or parties. Upon acceptance, the contract comes into effect.
2. Intention: Intention must be shown by all parties to enter into a binding contract.
3. Capacity: All parties to the contract must be legally capable. For example, minors and persons of unsound mind may be excluded from certain types of contract.
4. Consent: Consent must be proper and not obtained by fraud or duress.
5. Legality: The contract must be formed within the boundaries of the law. For example, a contract to commit a crime would not be binding.
6. Possibility: Contracts formed to undertake impossible tasks are unenforceable.
7. Each party must contribute something in consideration of the other's promise. Consideration:
   - must be real
   - must be legal
   - need not necessarily be adequate
   - must be possible
   - must not be in the past
   - and must move from the promisee.

Privity

Privity is a legal doctrine which recognizes that only a party to a contract may sue upon it. There are certain exceptions to this general rule, e.g., where an agency relationship exists, the principal is bound by contracts entered into by his or her agent with third parties.

Discharge

A contract may be discharged by:

- performance, i.e., realization of the agreement within the terms of the contract;
- agreement by all parties to cease their contractual relationship;
- operation of law, e.g., if a contract is made for a limited period, and that period expires;
- frustration or subsequent impossibility. Performance of the contract may be possible at the outset, but later frustrated by events (e.g., death of a party, destruction of an element constituting the basis of the contract).

Breach

A breach occurs when a party to the contract does not fulfill obligations. If the breach "goes to the root" of the agreement, the contract is treated as discharged. Such breaches are referred to as "material," and the injured party may seek one of the following remedies:

1. Refusal of further performance of the contract.
2. Rescission: This is a discretionary remedy, enabling the courts to cancel or annul the contract.
3. An action for specific performance: If successful, the court orders the party in breach to fulfill his or her obligations under the contract.

4. An action for an injunction: An injunction is a legal means of preventing further action by the party in breach.
5. An action for damages: This is the most common remedy for breach of contract. Damages can be:
    - General, i.e., arising out of the breach
    - Nominal, if the breach is merely technical
    - Punitive or exemplary, if the court considers the defendant's behavior particularly deplorable
    - Liquidated, i.e., ascertained by an agreed method to assess damages (e.g., $x per day when the completion date is exceeded), as in most construction contracts.
    - Unliquidated, i.e., unascertained, or determined after injury occurs.
6. An action for a *quantum meruit*: This is a claim for an amount equal to that which the plaintiff has earned in respect of the contract.

BUILDING CONTRACTS

Types of Contract

A building contract may take any form which is agreeable to the parties involved, but certain proven types of contract have been developed which are useful for certain building projects. These are:

1. Fixed price/stipulated sum contracts
2. Cost-type contracts

Fixed Price/Stipulated Sum

This method of agreeing a contract price is widely used in the construction industry, where one party pays an acceptable sum for a specified amount of work to another party who agrees to undertake it. It is nearly always used in connection with competitively bid work (some public bodies are constrained by law to use this method), and has the advantages of:

- enabling the owner to know the final cost of construction at the outset of the work;
- releasing the contractor from having to keep accurate time records for the owner's scrutiny.

Stipulated sum contracts have certain disadvantages; for example, escalation or inflation of prices or unforeseen circumstances might affect the contractor's fixed profit margin. In some cases, this could mean a higher base bid to cover such contingencies, and so the owner may pay more than is strictly necessary. However, standard forms of contract often include equitable clauses to deal with these matters (e.g., escalation clauses,

concealed conditions clauses, etc.). The fixed price method of contracting is most suited to building projects of a predictable nature, where a full set of construction documents is available. In federal projects a fixed price-incentive firm method of contracting has been developed.

Cost-type Contracts

In this type of contract, the owner reimburses the contractor for the actual cost of completing the work, together with a negotiated fee. The fee might be:

- A percentage of the final construction cost
- A fixed fee
- An award fee

The method is useful:

- In negotiated contracts
- Where unknown conditions might be encountered
- If new building methods or materials are being used
- Where final scheme designs are not fully completed

It has the advantage of not needing a full set of documents before a price can be negotiated and work started, and enables the owner to bring the contractor into the process at an earlier stage, if required. It may be disadvantageous in that the owner is uncertain of the final cost and, in its percentage form, the cost-type contract gives no incentive to the contractor to keep cost down. Furthermore, the contractor will be obliged to keep accurate records of the work for payment purposes. In fact, cost-type contracts are prohibited in some states for certain kinds of work.

Certain variations upon the cost-type contract have been developed which make it more viable. These are:

a. Cost plus award fee
b. Cost plus incentive fee

Cost Plus Award Fee (CPAF Contract) This is often used in federal procurement work where the fee is negotiated on a percentage of the agreed estimated final cost. Added to this will then be an award fee, which may be two or three times the base fee, and is paid by the owner upon previously established criteria.

Cost Plus Incentive Fee (CPIF Contract) By this system, the contracting parties negotiate a target cost and fee, a base and ceiling fee, and a fee adjustment formula to provide an incentive for early and economic completion.

Foreign Contracts

Architects may often be called upon to work in other states or, increasingly, other countries. As well as considering licensing requirements (see page 31), the architect should take great care at the contract formation stage to avoid difficulties which might arise in enforcing agreements due to:

- conflicts between state laws (e.g., lien laws);
- conflicts between national laws;
- the contractual capacity or immunity of parties operating in other states or countries.

Before any major agreements with a foreign element are entered into, it would be prudent to check the legal position with legal counsel.

Contract Checklist

Some of the more important factors to be considered when determining the type of building contract are:

- Type of project (unusual building, renovation, etc.)
- Methods of construction proposed (experimental techniques, etc.)
- Size and complexity of project
- Time constraints
- Finance available
- Degree of certainty of the owner's requirements
- Progress of the construction documents
- Probability of further changes
- Amount of information available at contract formation
- Availability/desirability of accurate cost prediction
- Expertise necessary/available
- External factors or problems (e.g., site constraints, labor shortages, etc.)
- Quality of work required (luxury, low cost, etc.)

STANDARD FORMS OF CONTRACT

Just as any type of contract can be selected by the parties involved, so any form of agreement can be used to determine the terms and conditions of the contractual relationship.

However, in the interests of both parties, it is generally recognized that a format which has common usage and understanding is preferable. Standard forms of contract have been developed by a number of bodies, including:

- Professional associations
- The federal government
- State agencies
- Large institutions and private sector organizations

In some cases, house forms are a requirement of the owner. These should be carefully scrutinized, particularly in relation to the owner-architect agreement to ensure that provisions affecting the architect's duties do not violate state licensing requirements or increase the architect's liability.

Where possible, AIA contract documents should be recommended. These have been developed over a long period of time, and are recognized throughout the construction industry. Furthermore, they may be used in conjunction with a wide range of other AIA standardized documents.

Supplementary Conditions

Despite the broad and thorough coverage of the standard forms of contract, it is not unusual for variations or additions to be required. These may occur through:

- Owner's requirements
- The nature of the project
- Local/state legal requirements
- Climatic or physical factors

Where variations to the General Conditions are necessary or required, they may be included:

- In the Instructions to Bidders
- By the addition of the AIA Supplementary Conditions
- By inclusion in the General Requirements of the Specifications (Division I, Uniform Construction Index)
- In the owner-contractor agreement
- By modifications of the General Conditions of the Contract

Location of new addenda usually follows these generally accepted principles:

- Anything concerning the bidding phase and not affecting the construction phase should be included in the Instructions to Bidders.
- Matters going to the root of the contract (price, time, etc.) should be included in the owner-contractor agreement.
- Legal matters which may vary according to location (indemnification, insurance, etc.) should be dealt with in the Supplementary Conditions.
- Matters of a procedural or administrative nature (e.g., temporary structures, etc.) should be included in the General Requirements (Division I) of the Specifications.

Great care should be taken in the reformulation of contract documents if any changes are anticipated, and certain matters (e.g., legal responsibilities) should not be attempted without professional legal assistance.

CONTRACTOR SELECTION

During the development of each scheme, it will be necessary to establish on what basis the project will be constructed. This will have been reviewed with the owner much earlier in the process, in relation to other important factors, such as time or finance available, type of project, or site characteristics.

Once such variables have been assessed, certain alternative construction procedures can be considered. These might include:

- Single or separate contract systems
- Negotiated or competitively bid contracts
- Types of building contracts (see page 64)

Single Contracts

In most construction projects, a prime contractor is responsible for the full extent of the construction involved. If the work requires more labor or skill than the prime contractor can supply, subcontractors (and even sub-subcontractors) will be hired, but will remain the responsibility of the prime contractor in matters of liability to the owner, payment, etc.

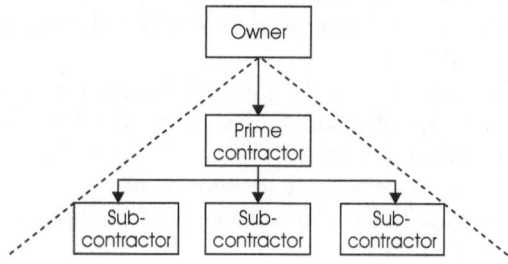

Figure 5.1

Separate Contracts

In some instances (e.g., certain state work and some larger projects), contractors will be selected for specific and distinct divisions of the work (electrical, mechanical, etc.). There is no prime contractor as such because all contractors will have an equal relationship with the owner. This system has the advantage of reducing the prime contractor's extra charge for administration of the

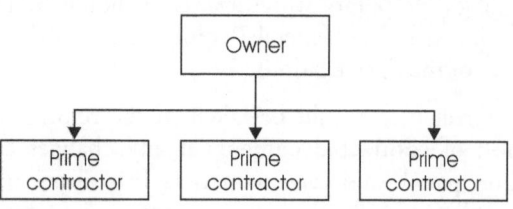

Figure 5.2

subcontractors and reduces the expense of double insurance. However, it may also complicate the relationship between the contractors involved. Since no hierarchy of responsibility exists between them, management and supervision of the project have to be coordinated carefully to ensure smooth and uninterrupted transfer between the individual work forces. Problems have been known to occur in matters of delay, cleaning up, etc., and the role of the construction manager (see page 23) is useful in this contracting system as a coordinator and supervisor of the various work forces. Alternatively, the architect could be employed by the owner to undertake this task.

Negotiated or Competitively Bid Contracts

The selection of the contractor can be undertaken in one of two ways, depending upon the character of the project:

Negotiated Contracts

The owner can select a contractor directly based upon the latter's reputation, recommendation, etc., and then negotiate the terms of agreement and form of payment. This may be appropriate:

- Where the contractor possesses skill or experience relevant to the project
- Where quality and not economy is a major determinant
- Where an early completion of the project is desired
- Where details of the final scheme are incomplete

Using the direct selection approach, the owner need not wait until the normal selection phase to begin construction. This means that completed drawings and specifications are not necessary for work to start, and that the contractor's skill and expertise may be brought into the design process.

However, there is no competition in this form of selection, making it potentially unsuitable where a low overall price is sought. Usually, a cost plus fee contract (see page 64) is used in conjunction with negotiated contracts.

Competitively Bid Contracts

In order to obtain the lowest possible price for the work, completed sets of contract documents are sent to a number of contractors who bid against each other. Usually, the lowest bidder is awarded the contract (see page 69).

Public agencies often require this method of contracting, which is best suited to projects of a straightforward, traditional nature, where no unforeseen problems are likely.

It is widely used in the construction industry, and necessitates complete design documentation to enable accurate bidding.

In the last few years, certain alternatives to the basic contracting methods have emerged. These include:

- Fast tracking
- Design-build work
- Turnkey contracts
- Bridging

Fast Tracking

As previously stated, if the contractor is selected on a negotiated basis, it is possible to begin construction work before the completion of the design phase. This method of overlapping the design and construction work is known as "fast tracking."

Design-Build Work

In the traditional model of building, the phases of design and construction are separate and usually undertaken by different specialists, and therefore different firms. It is increasingly common for companies to provide a package combining all the functions of the building process, often in large specialist-type projects, allowing the owner to contract with only one party to provide the complete building. This has the advantage of allowing fast tracking, and combining the skill of the designers and the constructors. It does, however, deprive the owner of an expert agent to look out for his or her best interests throughout the project.

The website of the Design-Build Institute of America carries further information on this growing mode of project delivery (www.dbia.org).

Turnkey Contracts

This kind of contract usually relates to projects where a developer proposes and constructs an entire development (including the purchase of the site) and hands it over to the owner ready for immediate occupancy when complete. It has been used in dealings with local housing authorities.

Bridging

Bridging is a variation of design-build wherein the owner hires an architect to prepare a preliminary building design and performance criteria. The concept and criteria information are then "bridged" to a design-build team that generates the construction contract documents and completes construction.

To facilitate the process of contractor selection, the AIA has developed AIA Document G612-2001, Owner's Instructions to the Architect Regarding the Construction Contract. This form may be filled out by the owner in advance of the selection stage, and can help to clarify the requirements and preferences for the architect so that appropriate action can be advised.

BIDDING

If the contractor is going to be selected by competitive means, certain procedures can be implemented to facilitate the task.

Selection of Bidders

The initial process of identifying possible bidders may be:

- Open
- Selective

Open Bidding

Where the maximum number of bidders is considered desirable (usually in public work), an advertisement to bid will be published in trade or governmental publications or professional journals, inviting any interested contractors to participate in the process.

Selective Bidding

If a limited number of bidders is preferred, an invitation to bid will be sent to a number of contractors. These will be singled out by reputation, recommendation, previous contact with either the owner or the architect, etc.

Contractor Qualification

Prospective bidders should be chosen for their ability to successfully undertake the project, and it may be necessary to establish their suitability before bidding documents are issued. In some cases, the contractor's reputation or relationship with the owner will be sufficient, but AIA Document A305, Contractor's Qualification Statement, may help to outline contractors' suitability.

The document when completed provides full details of the contractor's business record, and enables the owner and the architect to gain a clear impression of such details as:

- History of the business
- Organization and scope of operations
- Past record of construction work (type of work, range of experience, etc.)
- Trade and bond references
- Bonding company
- Details of assets and liabilities

The qualification statement can be used as a prequalification stage in the open bidding process to eliminate unsuitable bidders and cut down the administration involved in high numbers.

Once the bidders have been identified and contacted to ensure their interest, a package of information concerning the proposed project is issued. The package includes:

- The invitation/advertisement to bid
- Drawings and specifications (see page 41)
- The bid form
- Notice to Bidders
- Instructions to Bidders
- Proposed contract documents
- Bid security details (if required)

Drawings and Specifications

These documents, which should be as complete and unambiguous as possible to allow the contractors to bid accurately, are sent free of charge to the bidders. The number of sets necessary for each bidder varies, depending on project size and complexity. More sets may be required to expedite the bidding process; the architect can require additional payment for the extra work necessary to accomplish this. Similarly, if any of the bidders asks for extra copies, they may be provided at their expense.

To ensure return of the bidding documents by unsuccessful bidders, a deposit is usually required which is returned upon receipt of outstanding documents.

Notice to Bidders

This may be included in the bidding documents, and informs prospective bidders of their opportunity to bid, and conditions and requirements involved.

Instructions to Bidders

AIA Document A701, Instructions to Bidders, provides all relevant information concerning the detailed requirements involved in the bidding process, including:

- Definitions
- Bidding documents
- Consideration of bids
- Owner-contractor agreement
- Supplementary instructions
- Bidder's representations
- Bidding procedures
- Post-bid information
- Performance payment bonds (see page 74)

Contract Documents

All documents intended for use in the proposed project should be sent to each bidder for examination, including the conditions (e.g., AIA Document A201) and any other applicable addenda or Supplementary Conditions.

Bid Form

This form, which should be sent to all bidders, contains all relevant information concerning the project. Each bidder will then return the document complete with the price of the work, or base bid, and any other figures which may be appropriate (e.g., alternate bids, substitutions, etc.).

Bid Security

In order to ensure each bidder's commitment to their base bid, some form of security may be required by the owner, which should be submitted along with the returned bid form. The security might take the form of cash, a certified check, or a bid bond (AIA Document A301; see page 72). The bond could be expressed either in a lump sum or as a percentage of the base bid, although the former is usually preferred by bidders, as it does not reveal their bid before opening. The bond ensures that, in the event of the successful bidder refusing to undertake the work for the bid specified, the whole or part of the security may be forfeit. The amount of the penalty is usually determined as the difference between the selected bid and the next lowest.

Variations

Where possible, documentation necessary for accurate bidding should be comprehensive and unambiguous. In some instances, however, it may be necessary to provide some alternatives in the bidding process if requirements cannot be fully determined. Two mechanisms which allow this are:

- Alternates
- Unit prices

Alternates

An alternate bid may be required or accepted for a specific section of the work, and should be included in the calculation of the base bid. This procedure can be useful in helping to keep costs within a certain budget, but should be used sparingly and not employed to give one bidder preference over the others.

Unit Prices

Unit prices provide a means of measurement which can be included in the bid, indicating a price per unit for materials and/or services. It is useful in giving an idea of price calculation for unknown quantities or variable factors and, again, should be restricted in use if the overall budgetary figure needs to be controlled.

INSTRUCTIONS TO BIDDERS

In implementing the AIA procedures of contractor selection by requesting bids, certain rules have been developed which should be adhered to by all parties concerned. These are outlined in AIA Document A701, Instructions to Bidders.

The procedural format, following the mailing of necessary bidding documents includes:

- Modification of bidding documents
- Submission of bids
- Bid opening
- Selection
- Announcement
- Contract award

Modification of Bidding Documents

Certain queries or adjustments to the documents might be necessary or requested prior to the closing date for submission. These are usually in the form of:

1. Interpretations
2. Substitutions

Interpretations

If any of the bidders should discover errors or ambiguities in the documentation, they must inform the architect in writing at least seven days prior to the submission date. Any changes or addenda will then be issued by the architect to all bidders.

Substitutions

Should any of the bidders wish to substitute materials or services otherwise than specified in the bidding documents, the architect must receive a request for approval in writing at least ten days prior to the submission date. If the architect decides that the submission is acceptable, all parties will be notified by addendum, although no addenda can be made within four days of the final receipt date except a notice cancelling or postponing the request for bids.

Submission of Bids

Bids must be delivered in writing, contained in sealed, opaque envelopes prior to the time and date specified in the advertisement/invitation to bid. Oral bids are not acceptable. Any bids received after the specified time should be returned unopened.

Bid Opening

If the bids are opened in public, they are often read aloud, whereas if opened in private, the bidding information may be sent to all bidders at the owner's discretion. The owner need not accept any of the bids if they appear too high, and may reject any bid not in conformance with the stated requirements. The bidding documents do provide, however, that if a contractor is chosen, it will be on the basis of the lowest responsible bid. The decision is usually reached within ten days of the bid opening.

In publicly bid work, the owner is often constrained by law to accept the lowest responsible bidder, and may be held criminally liable if the selection does not conform to these requirements (i.e., the lowest monetary bid, coupled with the owner's satisfaction that the contractor can successfully undertake the work). In privately bid work, the commitment is not as clear, although the rules of bidding should be adhered to. Granting of the contract to any other than the lowest bidder should only be made with very good reason to prevent suspicion of favoritism, and ill feeling among the contractors.

Selection

At any time prior to the bid opening, all bidders may withdraw or modify their bids. However, once the bids are opened, the bidders cannot make changes or withdraw from the process for a period stipulated in the bidding documents (e.g., thirty days). Once selected, the successful bidder must undertake the work for the agreed price, or risk forfeiture of the bid bond (if any). Exceptions to this are sometimes made if the bidder can prove substantial error in the bid calculation, in which case withdrawal might be appropriate, with award of the contract to the next lowest bidder. Alternatively, the contract may be rebid. Defaulting bidders should be disqualified from any further bidding on the same project, and no bid correction should be permitted, except for minor clerical errors and alterations.

Announcement

When a contractor has been selected (usually within ten days of bid opening), all bidding parties should be informed of the decision. The unsuccessful bidders are often given a list of the bid figures, and the bid deposits are returned once the documentation is received. The successful bidder should be informed of the decision in a way which does not form a legally binding agreement prior to the signing of the contract documents. Usually, the bids of the next two or three lowest bidders will be retained for a period as a contingency measure.

At this stage, each party to the proposed building contract may provide further information and/or assurances to the other:

• The owner will, upon request, prove to the contractor that sufficient financial arrangements have been made to undertake the project.

• The contractor, within seven days of the contract award, should furnish:
  a. details of the amount of work to be undertaken by the contractor's forces;
  b. names of proposed suppliers of material and equipment;
  c. a list of intended subcontractors for the architect's approval (see page 77).

The contractor may also be asked for:

• A contractor's Qualification Statement (if appropriate and if not completed prior to selection)
• Proof of the responsibility and reliability of the work force
• Bonds, in accordance with the owner's requirements as expressed in the Instructions to Bidders

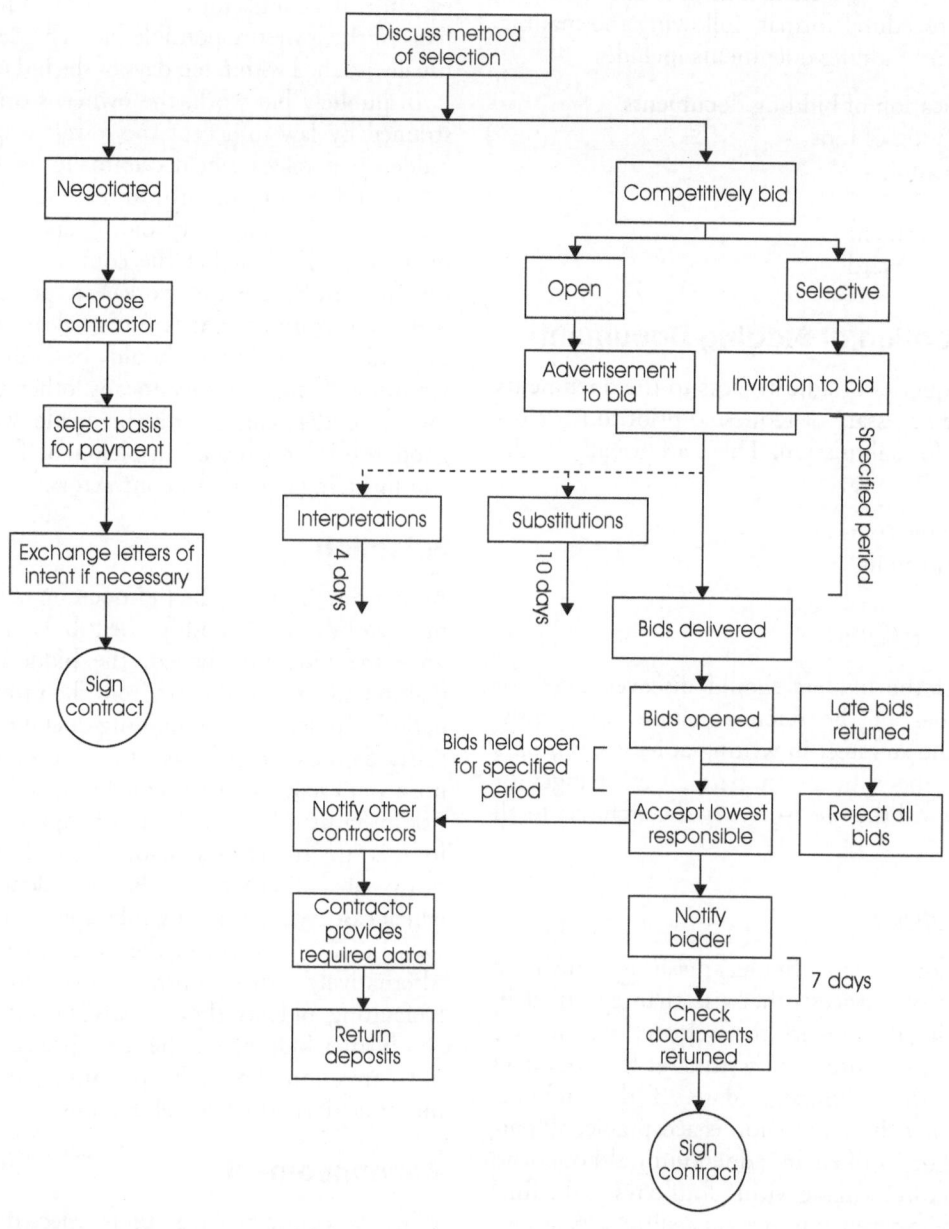

Figure 5.3

When these preliminary matters have been dealt with, and the contract documents are prepared, both parties will be ready to enter into the contractual agreement.

CONTRACT PROCEDURES

The contract documents comprise:

- The owner-contractor agreement
- The conditions of the contract (including any supplementary details or other conditions)
- The drawings
- The specifications
- Any addenda previously issued, or modifications (i.e., written amendments to the contract, signed by both parties, e.g., change orders, written interpretations, or minor changes)
- Related documents and agreements
- Performance bond and labor and material payment bond
- Owner's insurance and contractor's insurance

When the documents are ready, they should be sent to the parties for signing with a cover letter.

Notice to Proceed

This is written authorization from the owner to the contractor establishing a date of commencement and completion of the building work. The Notice to Proceed is used if the work is started after (not before) the date of the signing of the contract.

Letter of Intent

Should construction need to be started before the contract documents have been signed (e.g., where time is of the essence), a letter of intent may be sent by the owner, giving the contractor authority to proceed. If used, the letter should be carefully drafted to avoid any conflict with the actual contract documents, and legal assistance should be sought. The letter should emphasize that no subcontracts should be effected, nor should any materials be ordered other than those relevant to the specific work permitted. Insurance should be carefully considered if a letter of intent is used, and it should be made very clear that the letter will cease to have effect upon the signing of the actual contract.

Once the contractual relationship is established, certain obligations must be met by both parties, including:

- Owner capability (see page 70)
- Contractor's work schedule
- List of subcontractors
- Schedule of values
- Certification of insurance
- Permits

Contractor's Work Schedule

As soon as the contract has been awarded, the contractor should provide for the architect's information an estimated schedule of progress. This is usually in the form of:

- A bar chart
- Critical path method

Bar Chart

A bar chart indicates the work, divided by trades or operations, against which a time scale can be set. The progress of the work can be plotted between the two.

Critical Path Method

Critical path analysis is a project planning device which aims to optimize time and operations on site. The system divides various activities which are sequenced in terms of their interrelationship. When the time factor is added, a path may be plotted which reveals the most efficient operational procedures which should be followed. The schedule can be monitored by regular assessment of actual achievement on site. This enables continued prioritizing and adjustment throughout the period, to enable maximum efficiency in allotting time for the various stages of the project.

PERT (Project Evaluation Review Technique)

This is a method of scheduling which establishes, in chart form, activities and operations anticipated in the project layout which can introduce a cost element into the programming. PERT has not been commonly adopted in the construction industry.

List of Subcontractors

See page 77.

Schedule of Values

Prior to the first application for payment, the contractor must submit a Schedule of Values to the architect, together with any data supporting its accuracy that the architect may require. This then forms the basis for reviewing future applications for payment, and should indicate the sections of the contract sum provided for the various parts of the work.

THE AMERICAN INSTITUTE OF ARCHITECTS

AIA Document A310

Bid Bond

KNOW ALL MEN BY THESE PRESENTS, that we
(Here insert full name and address or legal title of Contractor)

as Principal, hereinafter called the Principal, and
(Here insert full name and address or legal title of Surety)

a corporation duly organized under the laws of the State of
as Surety, hereinafter called the Surety, are held and firmly bound unto
(Here insert full name and address or legal title of Owner)

as Obligee, hereinafter called the Obligee, in the sum of

Dollars ($),
for the payment of which sum well and truly to be made, the said Principal and the said Surety, bind ourselves, our heirs, executors, administrators, successors and assigns, jointly and severally, firmly by these presents.

WHEREAS, the Principal has submitted a bid for
(Here insert full name, address and description of project)

NOW, THEREFORE, if the Obligee shall accept the bid of the Principal and the Principal shall enter into a Contract with the Obligee in accordance with the terms of such bid, and give such bond or bonds as may be specified in the bidding or Contract Documents with good and sufficient surety for the faithful performance of such Contract and for the prompt payment of labor and material furnished in the prosecution thereof, or in the event of the failure of the Principal to enter such Contract and give such bond or bonds, if the Principal shall pay to the Obligee the difference not to exceed the penalty hereof between the amount specified in said bid and such larger amount for which the Obligee may in good faith contract with another party to perform the Work covered by said bid, then this obligation shall be null and void, otherwise to remain in full force and effect.

Signed and sealed this   day of   19

_____
(Witness)

(Principal)	(Seal)
(Title)	

_____
(Witness)

(Surety)	(Seal)
(Title)	

AIA DOCUMENT A310 • BID BOND • AIA ® • FEBRUARY 1970 ED • THE AMERICAN INSTITUTE OF ARCHITECTS, 1735 N.Y. AVE., N.W., WASHINGTON, D. C. 20006   **1**

 Printed on Recycled Paper   9/93

AIA Document A310: Bid Bond

The American Institute of Architects is pleased to provide this sample copy of an AIA Contract Document for educational purposes. Created with the consensus of contractors, attorneys, architects and engineers, the AIA Contract Documents represent over 110 years of legal precedent.

SUPPLEMENTAL ATTACHMENT FOR ACORD CERTIFICATE OF INSURANCE 25-S (7/90).

AIA DOCUMENT G715

(This document replaces AIA Document G705, Certificate of Insurance.)

PROJECT _____

INSURED _____

A. General Liability   **Yes No N/A**

   1.   Does the General Aggregate apply to this Project only?   ☐ ☐ ☐

   2.   Does this policy include coverage for:
      a.   Premises—Operations?   ☐ ☐ ☐
      b.   Explosion, Collapse and Underground Hazards?   ☐ ☐ ☐
      c.   Personal Injury Coverage?   ☐ ☐ ☐
      d.   Products Coverage?   ☐ ☐ ☐
      e.   Completed Operations?   ☐ ☐ ☐
      f.   Contractual Coverage for the Insured's obligations in A201?   ☐ ☐ ☐

   3.   If coverage is written on a claims-made basis, what is the:
      a.   Retroactive Date? _____
      b.   Extended Reporting Date? _____

B. Worker's Compensation

   1.   If the Insured is exempt from Worker's Compensation statutes, does the Insured carry the equivalent Voluntary Compensation coverage?   ☐ ☐ ☐

C. Final Payment Information

   1.   Is this certificate being furnished in connection with the Contractor's request for final payment in accordance with the requirements of Subparagraphs 9.10.2 and 11.1.3 of AIA Document A201, General Conditions of the Contract for Construction?   ☐ ☐ ☐

   2.   If so, and if the policy period extends beyond termination of the Contract for Construction, is Completed Operations coverage for this Project continued for the balance of the policy period?   ☐ ☐ ☐

D. Termination Provisions

   1.   Has each policy shown on the certificate and this Supplement been endorsed to provide the holder with 30 days notice of cancellation and/or expiration? List below any policies which do not contain this notice.   ☐ ☐ ☐

E. Other Provisions

_____
Authorized Representative

_____
Date of Issue

AIA CAUTION: You should sign an original AIA document which has this caution printed in red. An original assures that changes will not be obscured as may occur when documents are reproduced.

G715-1991   2

AIA Document G715: Supplemental Attachment

Certification of Insurance

The contractor should file with the owner (or with the architect if requested by the owner) certificates of insurance before starting work (AIA Document A201, Article 11.1.3).

Permits

Under the AIA General Conditions, the contractor is responsible for obtaining the building permit and certain other governmental requirements, e.g., licenses.

BONDS

Of the various measures often taken in the construction industry to minimize risk and potential loss, surety bonds are a common precaution. A surety bond is basically an assurance by one party which provides that specified obligations of another will be met, despite unforeseen or undesirable events. In reality, the cost of bonds, although technically borne by the contractor, is transferred to the owner by inclusion in the bid.

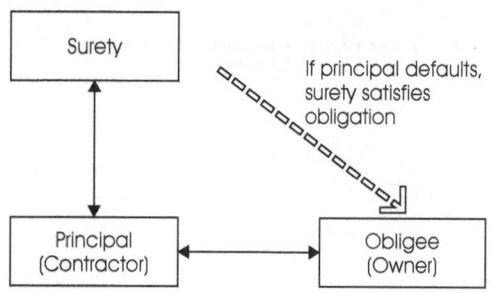

Figure 5.4

Types of Bond

Bonds used frequently in the construction industry are:

1. Bid bond
2. Performance bond

Bid Bond

(See page 72.) In order to ensure that the selected bidder signs the contract and fulfills other preliminary requirements, a bid bond may be requested. This should cover not less that 10 percent of the bid amount, and would be used to pay the owner the difference between the two lowest bids if the successful bidder decides to back out. The penalty for this cannot exceed the bond amount, which should be expressed as a specific sum, not a percentage of the bid.

Performance Bond

A performance bond ensures that all bids for labor and materials will not revert to the owner in the event of nonpayment by the contractor.

Combination bonds are considered inadvisable by the AIA as they can cause legal complications in the event of a claim. The AIA recommends the two-bond system as a preferable procedure. State laws should be checked regarding the use of bonds, as statutory requirements vary with regard to bond provisions.

If claims are made against bonds during a construction project, AIA Document B141, Owner-Architect Agreement, provides for additional payment to the architect for the work involved in making the necessary arrangements for the continuation of the project.

Other Bonds

Other forms of bonds sometimes used include:

- Payment bond
- License or permit bond
- Lien and no-lien bond
- Maintenance bond
- Release of retained percentage bond
- Statutory bond (check each state for requirements)
- Subcontract bond
- Termite bond

AIA Bonds

Although there are no standardized requirements for bonds compatible with all state laws and owner preference, the AIA produces certain forms which are helpful in many cases. These include:

- A310, Bid bond (see page 72)
- A312, Performance Bond and Payment Bond (see pages 75, 76)

In certain states, variations of the basic forms have been developed to comply with individual state laws for use in public and private construction projects.

In all matters relating to bonds and insurance the owner should seek expert advice. The architect should not attempt to provide this information, as it does not fall within architectural services and may be expressly proscribed by some professional liability insurance policies.

SUBCONTRACTORS AND SUPPLIERS

Under the single contract system, it is not unusual for prime contractors to sublet parts of the work

THE AMERICAN INSTITUTE OF ARCHITECTS

AIA Document A312

Performance Bond

Any singular reference to Contractor, Surety, Owner or other party shall be considered plural where applicable.

CONTRACTOR (Name and Address):

SURETY (Name and Principal Place of Business):

OWNER (Name and Address):

CONSTRUCTION CONTRACT
Date:
Amount:
Description (Name and Location):

BOND
Date (Not earlier than Construction Contract Date):
Amount:
Modifications to this Bond: ☐ None   ☐ See Page 3

CONTRACTOR AS PRINCIPAL   SURETY
Company: (Corporate Seal)   Company: (Corporate Seal)

Signature: _____   Signature: _____
Name and Title: Name and Title:

(Any additional signatures appear on page 3)

(FOR INFORMATION ONLY—Name, Address and Telephone)
AGENT or BROKER: OWNER'S REPRESENTATIVE (Architect, Engineer or
                                                 other party):

AIA DOCUMENT A312 • PERFORMANCE BOND AND PAYMENT BOND • DECEMBER 1984 ED. • AIA ®
THE AMERICAN INSTITUTE OF ARCHITECTS, 1735 NEW YORK AVE., N.W., WASHINGTON, D.C. 20006
THIRD PRINTING • MARCH 1987

A312-1984 1

AIA Document A312: Performance Bond and Payment Bond

THE AMERICAN INSTITUTE OF ARCHITECTS

AIA Document A312

Payment Bond

Any singular reference to Contractor, Surety, Owner or other party shall be considered plural where applicable.

CONTRACTOR (Name and Address):   SURETY (Name and Principal Place of Business):

OWNER (Name and Address):

CONSTRUCTION CONTRACT
  Date:
  Amount:
  Description (Name and Location):

BOND
  Date (Not earlier than Construction Contract Date):
  Amount:
  Modifications to this Bond: ☐ None   ☐ See Page 6

CONTRACTOR AS PRINCIPAL   SURETY
Company: (Corporate Seal)   Company: (Corporate Seal)

Signature: _____   Signature: _____
Name and Title: Name and Title:

(Any additional signatures appear on page 6)

(FOR INFORMATION ONLY—Name, Address and Telephone)
AGENT or BROKER: OWNER'S REPRESENTATIVE (Architect, Engineer or other party):

AIA DOCUMENT A312 • PERFORMANCE BOND AND PAYMENT BOND • DECEMBER 1984 ED. • AIA ®
THE AMERICAN INSTITUTE OF ARCHITECTS, 1735 NEW YORK AVE., N.W., WASHINGTON, D.C. 20006
THIRD PRINTING • MARCH 1987   **A312-1984 4**

AIA Document A312: Performance Bond and Payment Bond

to other contractors, either due to the size and scope of the project, or to take advantage of specialist skill or knowledge. If subcontracting is anticipated where AIA owner-contractor agreements are being used, a standard form of subcontract is advisable. The AIA produces Document A401, Contractor-Subcontractor Agreement Form, which can be used in conjunction with other AIA forms including A101, A107, A111, and A201.

The subcontractor agreement corresponds to the other AIA Documents in terms of:

- Responsibilities and liabilities
- Payment
- Relationships of parties

The subcontractor may, in turn, delegate responsibility to other contractors who are known as sub-subcontractors. The same relationship is established as with the contractor and subcontractor, although the prime contractor still retains overall responsibility for all work undertaken on site.

Selection

The contractor may select suitable subcontractors, and cannot be forced by the owner to work with anyone to which reasonable objection can be made. However, as soon as is practicable after the owner-contractor agreement is signed, the contractor should submit to the architect a list of proposed subcontractors and suppliers. The architect and/or owner may reasonably object to any of the names on the list, but such objection should be made promptly so that the contractor may submit a substitute. Objections should also be made on the basis of actual objections and accurate material to avoid potential claims of

defamation (see page 38). If the substitution is acceptable, the contract sum can be adjusted by Change Order (see pages 89 and 95) to accommodate any financial inequities caused by the change. No substitution of subcontractors should be made by the contractor without architect and/or owner knowledge and approval.

Payment

Payments to the subcontractor by the prime contractor are governed by the same requirements as the owner's payments to the contractor, although the AIA agreement provides for subcontractor payment within three working days of the owner's payment, reflecting the same retainage (see page 8).

The subcontractor may request information from the architect concerning the percentage of work completed or amounts certified (under the General Conditions 9.6.3), even though no contractual relationship exists between the two parties at any time. If a certificate of payment is withheld from the contractor through no fault of the subcontractor, the latter is nonetheless entitled to payment for work completed to date, and the contractor will be bound to pay it. If the subcontractor is not paid, Article 11.12.1 of AIA Document A401 states that, after giving due notice, the subcontractor can stop work until payment is made.

In some cases, payment made by the owner to the contractor may not reach the subcontractor (e.g., in the event of the contractor's bankruptcy). It is possible for the subcontractor to successfully claim against the property through the imposition of a lien (see page 8), causing the owner to pay twice for the same work. To avoid such problems,

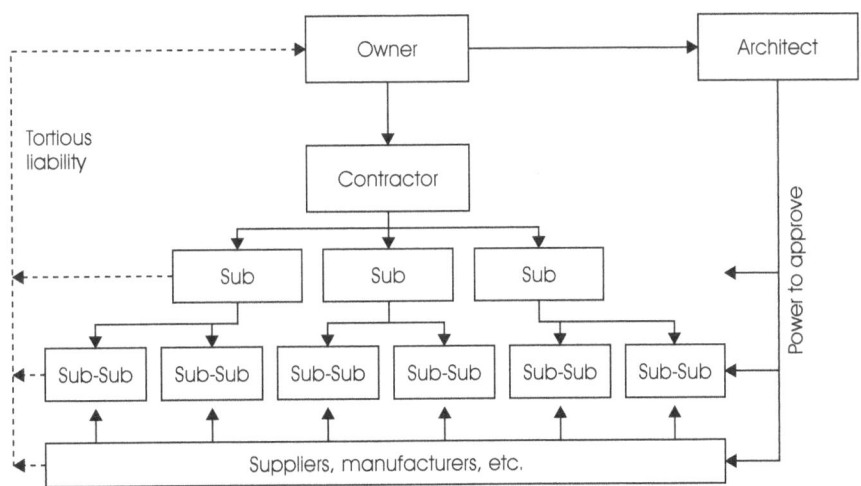

Figure 5.5

payment bonds should be used to prevent undue hardship to the owner (see page 76).

Suppliers

Material suppliers and manufacturers contract directly with the prime contractor, subcontractors, and sub-subcontractors and have no contractual relationship with either the owner or the architect. However, similar legal rights exist where nonpayment occurs, and appropriate bonds should be required from the contractor to give necessary protection.

The owner and the architect have similar rights under the AIA General Conditions A201 to reasonably object to any suppliers that the contractor intends to use.

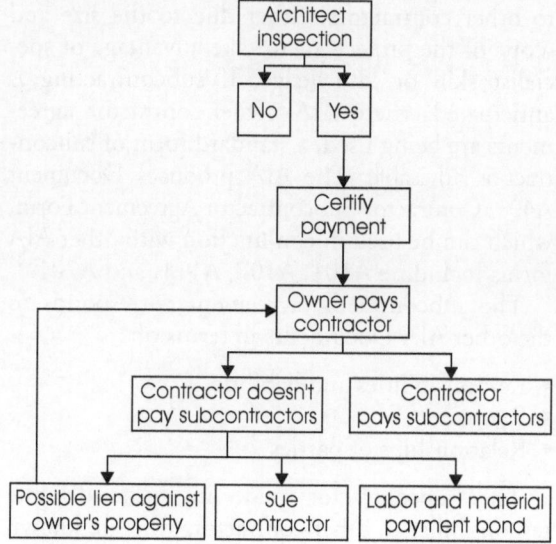

Figure 5.6

PRACTICE OVERVIEW

BIDDING AND SELECTION

Although much attention has been paid to the liability problems architects face during the design and construction phases, little consideration has been given to the interface between these phases: the selection of the contractor. Perhaps this is appropriate; after all, the architect usually receives approximately 5 percent of the fee for the bidding and negotiation phase, and research suggests that comparatively few cases involving architects originate from errors in this area. Still there are a few safeguards that architects can take to protect themselves from claims by clients, contractors, or subcontractors, and also to safeguard the owner's interests during the construction phase.

Complete Documentation

The more complete and accurate the drawings and specifications, the more precise the bids are likely to be. While some factors may make this difficult—a shortage of preparation time or uncertain owner requirements, for example—it is important to strive for unambiguous, accurate documentation, with a minimum of alternates or unit prices. In claims brought by contractors on the grounds of misinterpretation of the contract documents, the courts tend to find against the drafter (in this case the architect acting as the owner's agent) in matters of unclear contractual information. Alternatively, hazy documentation may lead a contractor to build a substantial contingency into the calculations, creating a base bid that far exceeds the architect's original projections. There has been a noticeable increase in claims against architects for inadequate prediction of construction costs. While cases vary, courts have found against architects where bids exceed the architect's estimates by 15 percent or less, a particularly worrying phenomenon since most malpractice insurance policies do not cover errors in cost estimating.

The Selection Process

Federal procurement procedures are remarkably specific, and state and local governmental rules are similarly designed to ensure fair and open competition. While such procedures are not required in the private sector, adherence to clearly articulated rules is still advisable, both to ensure a successful selection of a capable contractor at a fair price and to minimize the possibility of legal problems raised by irregular performance of one or more of the parties.

While some owners will insist upon their own procedures, the American Institute of Architects has developed widely accepted guidelines that should, whenever possible, be used. The procedures provide a series of orderly steps and safeguards that protect the owner and, by implication, the architect from unsuitable or unqualified contractors, while at the same time providing all bidders with an equitable basis for submitting their bids. Clear communication of all rules is very important at this stage. In instances where owners have not fully communicated their intentions and actions to bidders, legal action has ensued. For example, when an owner rejected all bids for a public project in Louisiana—a right established in the bid solicitation—one of the bidders filed suit when the former refused to provide reasons for the decision. The court ruled that the owner had not acted in good faith by failing to provide an explanation of the arbitrary action, contrary to the bidders' expectations.

Similarly, where an owner gave the contract to the second lowest bidder because, unlike the lowest, it was a local corporation, the court ruled that this was an invalid decision as the criteria for the final selection was not included in the information to bidders.

Selection Procedures

Most bidding procedures establish specific rules with regard to the bid opening. Strict adherence to these procedures is advisable, and care should be taken to avoid any collusion or conspiracy between the owner or the architect and one of the bidders. Exactly the same data should be sent to all bidders, including any clarification requested by one of the bidders prior to bid opening. Valid bids should, if possible, be opened in public and late bids should be returned unopened. The question of what constitutes a late bid has been the focus of a number of court cases instigated by disgruntled bidders who felt that a late bid gave a competitor an unfair edge. In one case involving a federal project, acceptance of a bid thirty seconds beyond the time of bid opening established in the bid solicitation was held to be invalid. While some public projects and certainly all privately bid work are likely to be less rigid on this point, it is advisable to reject late bids after the opening of the first one to prevent even the appearance of unfairness or competitive advantage.[1] While some flexibility may be considered acceptable under special circumstances— mail delivery problems, perhaps—the architects should advise the owner of potential problems that can occur whenever there is a departure from the established procedures.

The architect should also exercise great care in the advice given to the client about selection of the contractor. If it is believed that the contractor with the lowest bid should not be hired, the architect should articulate the reasons for a rejection with great care. Several suits have recently been brought against design professionals where they counseled against a particular contractor. In one case, an engineer advised against hiring the lowest bidder whom he felt did not possess enough experience to adequately complete the work. The bidder sued him for slander and interfering with a business relationship. The suit, however, was not successful, as it was held that the opinion was rendered in good faith.[2]

In another instance, the consultant, who was hired by a city to prepare specifications and help review the bids, was discovered to have "an unlawful relationship" with the contractor he recommended. The lowest bidder successfully sued the city, and the persons responsible pleaded guilty to criminal violations.[3] These and similar cases demonstrate the need to give advice on hiring only on an objective, factual basis, free of any conflict of interest and to record the process in writing. Documentation should be clear, concise, and well reasoned, avoiding any sweeping personal statements or colorful adjectives. In the absence of inaccuracy or perceived malice, slander will be very difficult to prove.

There is technically nothing to stop an owner from rejecting the lowest bid and hiring a pre-selected contractor who was encouraged to go through the competitive bidding process merely to keep the base bid down. However, architects should discourage this practice. Apart from the ethical implications of ignoring the rules, some outraged low bidders have sought legal relief to prevent the owner from proceeding with a project.

The selection of any other than the lowest responsible bidders is very difficult in publicly bid work, and great care should be taken in such cases to ensure that complete documentation can substantiate why a contractor was not considered "responsible" or the bid "responsive." The rationale should be based solely upon the criteria that were established in the bidding information.

In fact, some states insist upon open hearings to let disappointed bidders discuss the selection of anyone other than the lowest bidder in public projects.

Where the lowest bidder was considered insufficiently responsible based solely on rumors of poor performance, or where the owner had solicited insufficient data on a hitherto unknown contractor, courts have found against the owner for insubstantial reasoning. Owners are expected to consider a contractor's recent performance to verify if former problems have been eradicated.[4]

Summary

While contractor selection is a relatively minor duty by comparison to those in the design and construction phases, there are still ample opportunities for problems in what is, after all, a sensitive and highly competitive area. In the role as adviser to the owner, the architect should strive to ensure that the procedures adopted are initially sound and rigidly and fairly adhered to, particularly in publicly bid work, and that the owner is kept informed of the possible implications of straying from the established rules. This helps to protect the architect from claims by the owner for poor advice and from the contractor for collusion or slander. It furthermore helps to shield the owner from unforeseen problems and allows a smooth transition from the design to the construction phase, optimizing the chances for the successful completion of each project.

References

1. Jervis, B.M. and Levin, P., *Construction Law: Principles and Practice*, McGraw-Hill, 1988, p. 13.
2. *Riblet Tramway Co., Inc.* v. *Ericksen Associates, Inc.* 665F Supp. 81 (P.N.H. 1987).
3. *F. Buddie Contracting Inc.* v. *Seawright*, 595 F. Supp. 122 (N.D. Ohio 1984).
4. Jervis and Levin, *Construction Law*, p. 15.

Question & Answer

Once the client has picked a contractor, the architect's role in the selection is over, right?

In the traditional, competitively bid process, the contract is usually awarded to the lowest responsible bidder. The architect has an administrative role in the organization of the bidding process and advising the client on the outcome.

While the architect's compensation for this phase is comparatively small (typically around 5 percent of the overall amount), the architect plays an important role in overseeing the established bidding procedures, offering advice to the owner on the appropriate selection and reviewing the winning contractor's list of proposed subcontractors and suppliers (see page 77).

The architect has further responsibility in ensuring a smooth transition from the design to the construction phase through the signing of the building contract (see page 69.) and may have to further advise the client if there is a mistake in the winning bid. In this event, the contractor may be allowed to withdraw the winning bid or correct it to adjust for a legitimate mistake, although the architect should be careful if the corrected bid now exceeds the next lowest responsible bid. In publicly bid work, it may be necessary to re-bid the project, while in privately bid projects, the client should be advised accordingly on the potential for disgruntled contractors taking legal action.

The construction phase

AIA FORMS

Throughout the design and construction process, a number of procedures and operations involving the architect have to be carried out which require documentation and record. As a written format is always preferable, standardized forms are useful, and many larger organizations will prepare their own personalized paperwork. This includes letterheads, memorandum pads, and telephone message pads, but may also extend to more technical and detailed documents necessary in both office management and project administration.

The AIA produces a comprehensive collection of documents for use in the construction process which are strongly recommended for their generally accepted meaning, consistency of format, and interrelated content.

The AIA documents are divided into series reflecting different aspects of administration in architectural practice:

A Series: Owner/Contractor Documents

These include all agreements designed for various construction project types and their respective conditions of agreement, bond forms, and bidding-related documents.

B Series: Owner/Architect Documents

Standard forms of agreement, duties, responsibilities, and limitations of the authority of the architect's project representative, the owner/construction manager agreement, and the architect's qualification statement are published in this series.

C Series: Architect/Consultant Documents

These include forms of agreement between the architect and consultants, including the engineer, and joint venture forms.

D Series: Architect/Industry Documents

These cover procedures for calculating area and volume of buildings and a project checklist.

G Series: Construction Administration Documents

This series includes land survey requisitions, change orders, certificates of substantial completion, applications for payment, and several other formats developed to assist architects in the internal running of the practice and in dealings with specific projects.

The AIA also publishes the *Architect's Handbook of Professional Practice*, which is highly recommended for all practicing architects.

THE ARCHITECT'S DUTIES

Once the building contract has been signed, the architect's role in the construction process changes, together with the architect-owner relationship. During the design development phases, the architect is seen (by some courts of law) to fulfill the role of independent contractor, whereas during the construction stage, this role becomes that of a limited agent (see page 20). The limits of this role are expressed within the owner-architect and owner-contractor agreements, and great care should be taken by the architect not to exceed or mishandle the powers necessary for the administration of the contract. If the architect's powers are exceeded, such acts can be ratified by the owner, but it is obviously preferable to avoid the situation if possible.

The architect's duties can be loosely grouped into three categories:

- Performance evaluation
- Certification
- Adjustment

Performance Evaluation

A major part of the architect's work during the construction phase of a lump sum contract concerns ensuring that the work carried out conforms to the detail and quality required by the drawings and specifications. There are no powers granted to the architect which enable him or her to tell the contractor how to do the work, but certain provisions within the building contract enable the architect to provide a quality control measure on behalf of the owner. These provisions can be categorized as:

- Observation
- Inspection
- Approval

Access to Work

Contained within the provisions of the General Conditions are a number of clauses that enable the architect to undertake the required duties. The contractor agrees to allow the architect access to wherever the work is in progress, which includes workshops where components or fittings are constructed, as well as visits to the building site (AIA Document A201, Article 3.16.1).

Site Visits

The architect should visit the site at appropriate intervals to ensure that the work is compliant with the contract documents (AIA Document A201, Article 4.2.2). Frequency of visits will depend on a number of factors, including:

- Type of project
- Site conditions
- Complexity and size of project
- Stage of construction reached
- Type of owner
- Knowledge of the contractor
- Location of the site from the architect's office
- Whether an on-site architect or construction manager is being employed
- Whether additional fees are being charged for inspections
- Unforeseen events (e.g., bad weather)
- Specific events (e.g., covering-up)

On arrival at the site, the architect should report to the contractor or the named superintendent and should communicate project matters solely with that person for the duration of the site visit. A record should be kept of all visits, noting any observations, information supplied, and actions that should be taken.

In the normal progress of the work, site visits can arise either during, at the commencement, or at the completion of some of the following activities, depending upon the project:

- Establishment of datum points, bench marks, and building layout
- Dimensions and grade establishment
- Safety and security provisions
- Protection of trees or existing buildings
- Fences, hoardings, and signs
- Siting of storage areas
- Excavation and soil underfootings
- Public utility connections (telephone, gas, electricity, etc.)
- Foundations, reinforcements, and pile-driving
- Concrete tests, formwork, reinforcement, and pouring
- Structural frames
- Floor openings, sleeves and hangers, floor laying
- Quality and placing of concrete
- Weather precautions
- Masonry layout and materials
- Bonding and flashing
- Frames and prefabricated elements

- Partition layout, lathing, and drywalling
- Temporary enclosures, heat, light, and sanitation during site operations
- Protection of finished work
- Fittings and cabinetwork
- Tiling, electrical work, wiring, pipework, and installation of hardware and equipment
- Roofing installation
- Painting, varnishing, and surface finish
- Equipment/plumbing tests and inspections required by public authorities

Inspection

At certain stages during the construction process, the architect will appraise the work completed and issue a written judgment upon it. Appraisal is required for:

- Progress payments (AIA Document A201, Article 9.3.1, 9.6)
- Substantial completion (AIA Document A201, Articles 4.2.9, 9.8)
- Final Inspection (AIA Document A201, Article 9.10.1)

Approvals

In addition to the above inspections or in respect of other duties required under the building contract, the architect may be called upon to make certain judgments on aspects of the work in the form of an approval or rejection (AIA Document A201, Article 13.5). Such instances include:

- Tests and inspections (13.5)
- Uncovering of work (12.1)
- Approvals of samples and shop drawings (3.11, 3.12, 4.2.7)
- Schedule of values (9.2)
- Names of subcontractors (5.2.1)
- Supporting data for payments (9.4.2)

Certification

At the stages in the construction process where inspections are carried out, it is often necessary to certify approval in writing. Such approval has the effect of releasing payment to the contractor, and should be undertaken with the greatest of care and diligence. Certification may take the form of a letter sent to both the owner and the contractor, or one of the standard forms produced by the AIA specifically for the purpose (see page 96).

Payment

In certifying payment, the architect must be satisfied that the amount of payment represents the stated value of the work (which must be reasonably accurate), less the agreed retainage and less the total of earlier certificates. In addition, the architect should be sure there is nothing to prevent the certificate being granted (i.e., defective work not remedied: A201, Article 9.5.1) and require such evidence substantiating the contractor's right to payment as considered necessary (Article 9.3.1).

Adjustment

This category of architectural service is the least defined in the contract documents, although provisions for its implementation can be found throughout the General Conditions. Basically, the powers connected with this category ensure that, in the event of confusion or disagreement between the owner and contractor, or in the event of unforeseen changes or conditions occurring, the architect can act to maintain the continued progress and quality of the work. The architect's duties in this respect fall into two main categories:

- Interpretation
- Modification

Interpretation

Article 4.2.12 of the General Conditions gives the architect authority to render interpretations of the intent of the contract documents in the event of the parties failing to agree. This helps to solve any ambiguities, and to keep minor disagreements or unclear requirements from delaying the project. Either the owner or the contractor can require the architect to interpret an aspect of the contract documents, and the architect should make the decision in writing within a reasonable time (A201, Article 4.2.11).

In the role of interpreter, the architect is expected to act as arbitrator and "to secure faithful performance by both the Owner and Contractor." To help the architect assume an unbiased position in this, a quasi-arbitral immunity is granted for decisions made under this provision, removing any liability for the results of the interpretation, rendered in good faith by the architect. Consequently, the architect should undertake interpretive duties with a totally unbiased attitude, and not allow employment ties to the owner to affect the outcome of decisions.

Modification

Where circumstances or new requirements mean that the contract documents need to be amended,

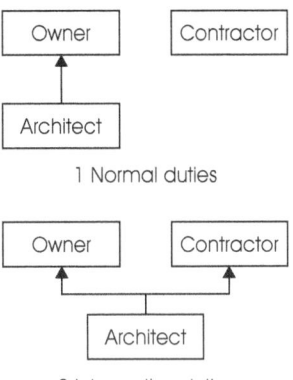

1 Normal duties

2 Interpretive duties

Figure 6.1

the architect is empowered to make certain minor changes (Article 7.4.1, see page 92) or issue (but not approve) Change Orders permitting new work to be undertaken (Article 7.2.1). Other actions in this respect may be carried out if prompted by the acts or omissions of either party (e.g., acceptance of nonconforming work by the owner, Article 13.3.1).

Authority to Reject Work

Work that does not conform with the contract documents may be rejected (Article 4.2.6).

PROGRESS APPRAISAL

A proportion of the architect's duties during the construction phase concerns the checking of the work to ensure that it will be completed by the agreed date of completion.

Several mechanisms may be used by the architect to monitor building progress throughout the project. These include:

- Site visits and reports
- The contractor's work schedule
- Schedule of values
- Meetings

Site Visits

Visits to the building site should be made at intervals appropriate to the stage of construction to familiarize the architect with the progress and quality of the work. Observations made during these visits should be recorded and copies sent to the parties involved, who may include:

- The owner
- Consultants

- Field architect, if appropriate
- Construction manager, if appropriate

Although any type of record will be sufficient for noting the outcome of site visits, standardized formats are recommended for the sake of consistency and conformity of files. The AIA publishes Document G711, Architect's Field Report (see page 90), which provides categories to note and comment upon the following:

- The stage of completion
- Temperature, weather
- Date
- Work in progress
- Persons present
- Conformance with schedule
- Any observations
- Items to verify
- Information or action required
- Photographic or video material may also provide a useful record of the stage of completion

The Contractor's Work Schedule

(See page 71.) This schedule represents the contractor's intended plan of work established at the outset of the construction phase. Comparison between the projected progress and actual advancement of the work provides a means of assessing the overall conformity of the project to the original timetable.

Schedule of Values

Similarly, the Schedule of Values which allocates value to various amounts or portions of the work can be used to a lesser degree to establish how well the original estimates of cost allocation match up to actual certification.

Meetings

Meetings between various parties concerned with the construction process may be held periodically. Types of meetings include:

- Practice meetings
- Contractor meetings
- Site meetings

Practice Meetings

Meetings between partners and/or employees may be held at intervals to discuss practice policy or a specific project that is in progress or scheduled to begin.

Contractor Meetings

The contractor and representatives may wish to meet with the subcontractors at intervals to discuss coordination of work on site. The architect or construction manager may be invited to attend where relevant.

Site Meetings

Meetings between parties representing different elements of the construction process could be necessary at intervals throughout the project.
   They may be held:

- At regular intervals
- At specific times during the construction process
- When problems occur
- When it seems necessary to provide an impetus

Those attending, in addition to the architect and the contractor (and/or the project representative) might include:

- The owner
- The construction manager (if one is employed)
- Consultant
- Subcontractors
- Others (e.g., the building inspector)

However, the nature of the project will largely determine the makeup and nature of the meeting.

Procedures

Although there is no standardized format involved in setting up and running meetings, certain basic guidelines are suggested for adoption.
   Whoever takes responsibility for chairing the meeting—and this role may be taken by the construction manager, the owner's representative, the contractor, or the architect—should prepare and distribute the minutes of the previous meeting and notify parties of the next one.
   Parties should be notified well in advance of the time and date of the proposed meeting, and be sent a copy of the previous meeting's minutes for consideration and filing. Parties unable to attend should notify the chair of their situations as soon as possible so that, in the event of their presence being necessary at the meeting, a new date may be scheduled which is amenable to all concerned.

The Agenda

The agenda of a typical site meeting might be set out in the following way.

The chair should:

- Call the meeting to order
- Take the names of those present
- Give the names of those sending apologies for their absence

The rest of the meeting might include:

- Agreeing the minutes of the last meeting, or dealing with any problems arising from them
- The architect's report
- The construction manager's report
- The contractor's report
- Any consultants' reports
- Discussion of project progress
- Procedures and communications necessary (any actions required, by whom, etc.)
- Any other business
- Time and place of next meeting

CONTRACT CHANGES

During the construction phase, it may become necessary to amend the original contract documents with addition, alteration, or deletion as a result of:

- Unforeseen or unexpected events
- New requirements
- New circumstances invalidating parts of the contract documents

The AIA General Conditions provide for changes to be made, but care should be taken to identify the nature of the change sought, and deal with it in the appropriate manner. Changes may fall into the following categories:

- A modification
- A cardinal change
- A constructive change
- A change
- A minor change
- Other forms of change

Modifications

At any time during the contract period, the owner and contractor may mutually agree to change the intent or substance of the contract between them. As the contract is a voluntary agreement between the parties, any joint acquiescence as to its content is acceptable, but great care should be taken in the modification of documents and the revised provisions for payment, work definition, etc.

Cardinal Changes

If the owner demands a change in the contract documents which goes beyond the intent of the original contract, this may be construed as a major change or, to use the federal procurement expression, a "cardinal change." Such a change may give the contractor sufficient justification to stop work and to claim damages for the owner's breach of contract. In privately funded projects, the contractor may wish to renegotiate payment for a new contract, whereas in publicly sponsored work, it may be necessary to re-advertise the project.

Constructive Changes

"Constructive changes" are referred to in federal procurement projects, and occur when the contractor is asked to undertake work:

a. which is different from that required by the contract;
b. which speeds up the project;
c. which requires added expenditures as a result of incorrect specifications.

If forced to make a constructive change, the contractor may require the contract sum to be adjusted accordingly.

Changes

If the AIA contract is used, as long as changes required by the owner are within the general scope of the contract, the contractor will be required to undertake the work, with or without the latter's consent. The contract time and the contract sum may be adjusted to compensate for the extra work. Payment for changed requirements could be:

- By mutual agreement on a lump sum
- By unit prices (either agreed upon, or previously stated in the contract documents)
- By an agreed cost of the work plus a fixed or percentage fee
- By determination of the architect

Ordering a Change

In the event that time is of the essence and there is absence of agreement on the terms of a Change Order, prior to requiring a change, it is often advisable to establish the final cost of the work involved. AIA Document G709, Proposal Request, may be sent to the contractor to ask for an account of the increased cost and/or time that will be necessary. If the owner decides to continue with the changed requirements, the architect will prepare and sign a Change Order (see page 95) and send it to the owner for signing before passing it on to the contractor.

Under the AIA General Conditions, the Change Order (AIA Document G701) is the only

The American Institute of Architects is pleased to provide this sample copy of an AIA Contract Document for educational purposes. Created with the consensus of contractors, attorneys, architects and engineers, the AIA Contract Documents represent over 110 years of legal precedent.

ARCHITECT'S
FIELD REPORT

OWNER ☐
ARCHITECT ☐
CONSULTANT ☐
FIELD ☐

AIA DOCUMENT G711

PROJECT:

CONTRACT:

FIELD REPORT NO:

ARCHITECT'S PROJECT NO:

DATE | TIME | WEATHER | TEMP. RANGE

EST. % OF COMPLETION | CONFORMANCE WITH SCHEDULE (+, −)

WORK IN PROGRESS | PRESENT AT SITE

OBSERVATIONS

ITEMS TO VERIFY

INFORMATION OR ACTION REQUIRED

ATTACHMENTS

REPORT BY:

AIA DOCUMENT G711 • ARCHITECT'S FIELD REPORT • OCTOBER 1972 EDITION • AIA® • © 1972
THE AMERICAN INSTITUTE OF ARCHITECTS, 1735 NEW YORK AVE., NW, WASHINGTON, D.C. 20006

page

AIA Document G711: Architect's Field Report

▓AIA® Document G710™ – 1992

Architect's Supplemental Instructions

OWNER	
ARCHITECT	
CONSULTANT	
CONTRACTOR	
FIELD	
OTHER	

PROJECT *(Name and address):*

ARCHITECT'S SUPPLEMENTAL INSTRUCTION NO:

OWNER *(Name and address):*

DATE OF ISSUANCE:

CONTRACT FOR:

FROM ARCHITECT *(Name and address):*

CONTRACT DATE:

TO CONTRACTOR *(Name and address):*

ARCHITECT'S PROJECT NUMBER:

The Work shall be carried out in accordance with the following supplemental instructions issued in accordance with the Contract Documents without change in Contract Sum or Contract Time. Proceeding with the Work in accordance with these instructions indicates your acknowledgment that there will be no change in the Contract Sum or Contract Time.

DESCRIPTION:

ATTACHMENTS:
(Here insert listing of documents that support description.)

ISSUED BY THE ARCHITECT:

_____   _____
(Signature)   *(Printed name and title)*

AIA Document G710: Architect's Supplemental Instructions

acceptable means by which the contract time or the contract sum may be altered. When signing the Change Order, the contractor indicates agreement with the proposed changes and becomes entitled to any justifiable extra payment (see page 89).

In the event that time is of the essence in the contract, and to prevent delay due to the administrative procedures involved, the process may be expedited by use of a Construction Change Directive (A201, Article 7.3). This is not a Change Order, but an authorization to the contractor to proceed with the work prior to the issuance of the Change Order.

Minor Changes

When alterations to the contract documents are considered necessary, but are sufficiently small as not to change the contract time or the contract sum, the architect is empowered to order such alterations which are referred to as "minor changes" (AIA Document A201, Article 7.4.1). Both the owner and the contractor will be bound by such written orders which are usually issued on AIA Document G710, Architect's Supplemental Instructions (see page 91).

Other Forms of Change

Due to the unpredictable and complex nature of many building projects, certain changes are sometimes necessary to provide for specific contingencies which include:

- Emergencies (AIA Document A201, Article 4.3.5)
- Concealed conditions (AIA Document A201, Article 4.3.4)
- Escalation and fluctuation of pricing

Emergencies

If the safety of persons or property is threatened in any way, the contractor may act at his or her discretion to prevent loss or injury. The architect may then determine the effects of the emergency on the project, and reflect them in a subsequent change order regarding contract time and contract sum.

Concealed Conditions

Because of the unpredictable nature of subsurface conditions, some contracts provide remedies to equitably adjust the contract sum and time if conditions prove to be materially different from those anticipated. The AIA General Conditions contain such provision, requiring that claims in respect of concealed conditions by either party be made within twenty-one days of their discovery.

Escalation and Fluctuation

Inflation and price escalation may make the estimation of a stipulated sum price difficult, possibly causing the contractor to overbid to protect against financial loss by erosion of profit. It is possible to add to any contract a fluctuations clause which provides an agreed method of calculation in the event of sudden price variations.

TIME AND DELAYS

Many stipulated sum building contracts are drafted on the basis that time is an important factor. The AIA General Conditions, for example, are drafted to include the provision that "time is of the essence" (Article 8.2.1). Such provisions make it important for the contractor to complete the work in conformance with the contract documents, on or before the date of substantial completion stipulated in the contract.

If the contractor fails to finish within the specified time, the contract is breached and several mechanisms may come into effect as a result, such as:

- Liquidated damages
- Termination
- Refusal of further payment
- Variations
- Extensions of time

Liquidated Damages

These basically represent a pre-agreed formula that can be used as a basis of penalty against the contractor for late work. They are usually determined as a fixed sum per day, payable for every working day beyond the date of substantial completion.

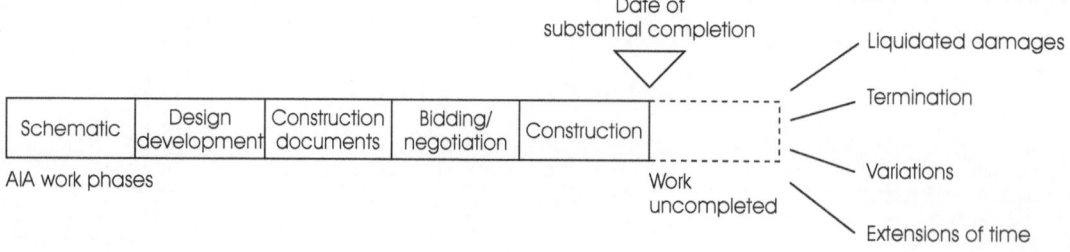

Figure 6.2

The extent of the financial amount involved for each day's delay will depend upon how critical prompt completion is considered by the owner.

The requirement is generally expressed in the bidding documents, and may affect the contractor's bid. If very high penalty clauses are used, they are often supplemented with a corresponding bonus clause. The bonus is often expressed as the same amount as the penalty and is payable to the contractor for every day saved prior to the Substantial Completion date.

Termination

In certain extreme circumstances of delay, there may be justification to terminate the contract between the owner and the contractor (see page 115).

Refusal of Further Payment

In some cases the contractor may be denied further payment. However, this should be handled carefully, as it may provide grounds for termination on the part of the contractor.

Note: Delay on the part of the contractor need not necessarily result in penalty if sufficient cause can be shown to substantiate a legitimate alteration to the contract documents.

Variations

If changes are made to the contract requirements by the owner, the architect can issue a Change Order which may provide for extra payment to the contractor, as well as extra time for completion (see page 89).

Extensions of Time

In some cases, unforeseen or unavoidable occurrences will delay the progress of the work through no fault of the contractor. The AIA General Conditions provide for extensions to the contract time to be granted by the architect for delays caused by:

- Act or neglect of the owner or architect (or employee of either)
- Act or neglect of a separate contractor (but not subcontractor: see page 77)
- Changes ordered to the work
- Labor disputes
- Fire
- Unusual delay in transportation
- Adverse weather conditions (not reasonably anticipatable)
- Unavoidable casualties
- Any cause beyond the contractor's control (or acts of God, including: earthquake, landslide, hurricane, tornado, lightning, flood, etc.)
- Delays authorized by the owner pending arbitration

- Any other cause that the architect determines to be justifiable

Claims for Extensions

Should the contractor feel that an extension is warranted, an application must be made in writing to the architect within twenty days of discovery of the event likely to cause delay. An indication of the probable effect of the delay upon the construction work should be included, and the contractor should be encouraged to make every reasonable effort to minimize the impact of the event on the general progress of the project.

The granting of an extension is not automatic; for example, bad weather alone may be insufficient to warrant extra time. It must be shown (e.g., by reference to meteorological records) that the weather in question was far worse than the norm for the year, and actually delayed operations on site.

Similarly, a claim for delay due to labor disputes may be disallowed if the dispute was in progress at the time of contract formation. The onus, therefore, is on the contractor to show both the justifiable reason for an extension, and its impact upon the progress of the work.

If there is sufficient cause to justify the granting of an extension, there may also be grounds for additional compensation. Such claims often arise from owner delay and decision, and may include:

- Site not ready in time for contractor occupation
- Delays in progress payments
- Delays in issuing change orders
- Delays in approving submittals
- Errors in drawings and/or specifications
- Administrative delays (poor coordination of separate contractors, inspection delays, etc.)

Claims can be made by the contractor to compensate for:

- Labor costs (including subcontractors' costs, wages, overtime, insurance, etc.)
- Equipment costs
- Material costs (additional and escalation costs)
- Overhead (field and office)
- Insurance and bond costs
- Other losses (seasonal problems, congestion on site, etc.)

Other factors relating to the time elements of the construction process include:

- Acceleration
- Stopping the work
- Impossibility

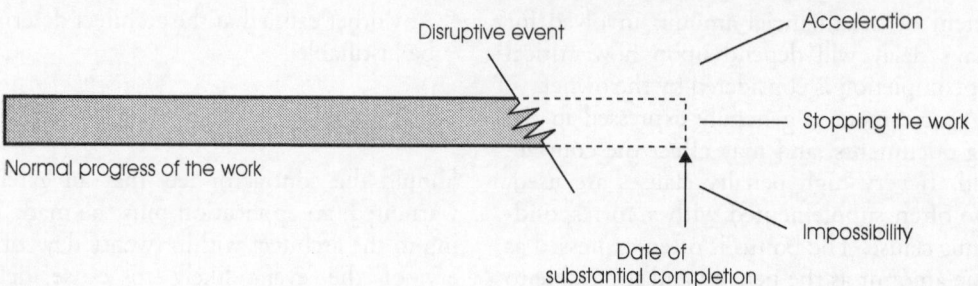

Figure 6.3

Acceleration

This can be defined in two ways:

1. Actual acceleration
2. Constructive acceleration

Actual Acceleration

This may take place if the contractor is requested to complete the work before the date established in the contract documents. Actual acceleration is at the contractor's discretion and may provide the basis for increased costs.

Constructive Acceleration

This is less clear in its definition, and may occur where the contractor has experienced delay, but has not been granted an extension of time. Consequently, the contractor must make up the lost time in order to finish by the agreed date, and effectively accelerate the pace of the work.

Although acceleration was essentially a federal procurement matter, it has become more applicable to private construction contracts.

Stopping the Work

Under certain circumstances, the work may be stopped, as opposed to delayed. This may be as a result of:

- Owner's instruction
- Circumstances forcing a construction suspension

Owner's Right to Stop the Work

AIA General Conditions (Article 2.3.1) provide the owner with the right to stop the work:

a. If the contractor fails to correct defective work
b. If the contractor persistently fails to carry out the work in conformance with the contract documents

The architect does not have the power to stop the work unless expressly authorized to do so by the owner in writing. Previous editions of the AIA contract included the architect's right to stop the work, but this was discontinued in the 1997 revision of the AIA Documents.

Suspension of Construction

This may take place:

- If the contractor is not paid
- If notices and/or information are delayed excessively
- If change orders are delayed
- If certificates are unreasonably withheld or delayed
- If the construction documents are defective

Constructive suspension allows the contractor to stop work (AIA Document A201, Article 9.7.1), and to claim extra compensation for the costs involved in shut-down, delay, and recommencement. Some suspensions may eventually lead to the termination of the construction contract by either party (see page 115).

Impossibility

A further reason which may be claimed as the cause of delay, and possibly lead to the termination of the contract, is impossibility of completion. If sufficient cause exists to prove that the work cannot be finished, the contractor may be excused from further performance, and might be able to recover damages from the owner.

Impossibility of completion is generally classified as either:

1. Actual
2. Practical

Actual Impossibility

This arises when events occur which actually prevent performance from taking place (e.g., acts of God, or determination by a governmental department).

Practical Impossibility

In this case, completion of the work is technically possible, but only at excessive cost due to subsequent events making the original contract sum inadequate. If it is unreasonable for the contractor to assume the higher costs, or where it would cause excessive difficulty, loss, or possible damage, the parties may be released from their contractual obligations.

▲AIA® Document G701™ – 2001

Change Order

PROJECT: *(Name and address)*

CHANGE ORDER NUMBER:

DATE:

ARCHITECT'S PROJECT NUMBER:

TO CONTRACTOR: *(Name and address)*

CONTRACT DATE:

CONTRACT FOR:

OWNER ☐
ARCHITECT ☐
CONTRACTOR ☐
FIELD ☐
OTHER ☐

The Contract is changed as follows:
(Include, where applicable, any undisputed amount attributable to previously executed Construction Change Directives)

The original (Contract Sum) (Guaranteed Maximum Price) was $ _____

The net change by previously authorized Change Orders $ _____

The (Contract Sum) (Guaranteed Maximum Price) prior to this Change Order was $ _____

The (Contract Sum) (Guaranteed Maximum Price) will be (increased) (decreased) (unchanged)

by this Change Order in the amount of $ _____

The new (Contract Sum) (Guaranteed Maximum Price) including this Change Order will be $ _____

The Contract Time will be (increased) (decreased) (unchanged) by   (   ) days

The date of Substantial Completion as of the date of this Change Order therefore is

(Note: This Change Order does not include changes in the Contract Sum, Contract Time or Guaranteed Maximum Price which have been authorized by Construction Change Directive until the cost and time have been agreed upon by both the Owner and Contractor, in which case a Change Order is executed to supersede the Construction Change Directive.)

NOT VALID UNTIL SIGNED BY THE ARCHITECT, CONTRACTOR AND OWNER.

ARCHITECT *(Firm name)*	CONTRACTOR *(Firm name)*	OWNER *(Firm name)*
ADDRESS	ADDRESS	ADDRESS
BY *(Signature)*	BY *(Signature)*	BY *(Signature)*
(Typed name)	*(Typed name)*	*(Typed name)*
DATE	DATE	DATE

CAUTION: You should sign an original AIA Contract Document, on which this text appears in RED. An original assures that changes will not be obscured.

▲AIA® Document G702™ – 1992

Application and Certificate for Payment

TO OWNER:

PROJECT:

APPLICATION NO:

PERIOD TO:

CONTRACT FOR:

FROM CONTRACTOR:

VIA ARCHITECT:

CONTRACT DATE:

PROJECT NOS:

Distribution to:

OWNER ☐
ARCHITECT ☐
CONTRACTOR ☐
FIELD ☐
OTHER ☐

CONTRACTOR'S APPLICATION FOR PAYMENT

Application is made for payment, as shown below, in connection with the Contract. Continuation Sheet, AIA Document G703, is attached.

1. ORIGINAL CONTRACT SUM .... $

2. Net change by Change Orders .... $

3. CONTRACT SUM TO DATE (Line 1 ± 2) .... $

4. TOTAL COMPLETED & STORED TO DATE (Column G on G703) ...... $

5. RETAINAGE:

   a. _____ % of Completed Work
      (Column D + E on G703) $

   b. _____ % of Stored Material
      (Column F on G703) $

   Total Retainage (Lines 5a + 5b or Total in Column I of G703)...... $

6. TOTAL EARNED LESS RETAINAGE .... $
   (Line 4 Less Line 5 Total)

7. LESS PREVIOUS CERTIFICATES FOR PAYMENT .... $
   (Line 6 from prior Certificate)

8. CURRENT PAYMENT DUE .... $

9. BALANCE TO FINISH, INCLUDING RETAINAGE
   (Line 3 less Line 6) $

CHANGE ORDER SUMMARY	ADDITIONS	DEDUCTIONS
Total changes approved in previous months by Owner	$	$
Total approved this Month	$	$
TOTALS	$	$
NET CHANGES by Change Order	$	

The undersigned Contractor certifies that to the best of the Contractor's knowledge, information and belief the Work covered by this Application for Payment has been completed in accordance with the Contract Documents, that all amounts have been paid by the Contractor for Work for which previous Certificates for Payment were issued and payments received from the Owner, and that current payment shown herein is now due.

CONTRACTOR:

By: _____ Date: _____

State of:
County of:
Subscribed and sworn to before
me this _____ day of _____
Notary Public:
My Commission expires:

ARCHITECT'S CERTIFICATE FOR PAYMENT

In accordance with the Contract Documents, based on on-site observations and the data comprising this application, the Architect certifies to the Owner that to the best of the Architect's knowledge, information and belief the Work has progressed as indicated, the quality of the Work is in accordance with the Contract Documents, and the Contractor is entitled to payment of the AMOUNT CERTIFIED.

AMOUNT CERTIFIED .... $ _____
(Attach explanation if amount certified differs from the amount applied. Initial all figures on this Application and on the Continuation Sheet that are changed to conform with the amount certified.)

ARCHITECT:

By: _____ Date: _____

This Certificate is not negotiable. The AMOUNT CERTIFIED is payable only to the Contractor named herein. Issuance, payment and acceptance of payment are without prejudice to any rights of the Owner or Contractor under this Contract

AIA Document G702: Application and Certificate for Payment

PRACTICE OVERVIEW

WHEN LEGAL LIABILITY LOOMS—WHAT TO DO FIRST

> After reading reports of many such cases one is forced to the conclusion, that with few exceptions, those who find themselves at law are the stupid, the negligent, the dishonest and the unreasonable. The average architect, endowed with honesty and a fair degree of skill … is not likely to become involved in litigation.
>
> (*The AIA Handbook of Architectural Practice*, 1923)

These comforting words from the 1923 *AIA Handbook of Professional Practice* reflect a world long since gone. As every architect is uncomfortably aware, the threat of legal liability is an unfortunate but ever-present part of modern-day practice. While that threat has diminished since its high point in the 1980s, there is still enough evidence of legal action to keep architects awake at night worrying about the angry call or the hand-delivered subpoena.

Of course, attorneys are there to help out but they may not always be necessary and can be expensive. Here then is a six-point guide for dealing with that nasty moment when legal action is threatened. It will not remove the threat, but may help you to work through the initial stages more rationally, saving time, energy and money in the process.

1. Don't Panic

Received a nasty letter from a former client, his or her lawyer or, even worse, a subpoena, threatening legal action? It's not a pleasant feeling and your initial urge will be to try and fix things as soon as possible—make a call, set up a meeting, sort things out. Be careful as that may be the intent of the communication—to make you jump into a hasty course of action that you may later regret. While it's not pleasant waiting for a dispute to be resolved, there is no hurry, despite any hectoring demands that may be made of you. Dispute resolution is a time-intensive business, and even the supposedly faster process of arbitration can take months (or longer) to conclude.

Resist the temptation to rush into action. Acknowledge correspondence by all means, but be prepared to go slowly and thoroughly at this stage.

2. Check the Files

Both as a means of reassurance and to some degree as a displacement activity, thoroughly check the documentation involved in the dispute. Pore over the drawings, read the letters and contract documents to see if there are any inherent problems—design errors, procedural irregularities, etc. If nothing else, the process will reacquaint you with the details of the project in question and hopefully reassure you that there are no really big errors in the files.

Of course, this process will be made immeasurably easier if you have been keeping excellent records. A professional "paper trail" of each project, including records of telephone calls, E-mails, letters, forms and contracts—anything in fact that affects time, money and the nature of the work—gives great credibility to the design professional in any case. As a matter of course, always look over your own shoulder during project procedures and wonder how what you are doing would look to an uninvolved third party several years from now—the actual situation should a court case emerge. Memories of the details of complex events fade over time and perceptions can vary widely, even about something as simple as a face-to-face conversation, so commit everything to writing and store the files well for at least the length of the statute of limitation or repose established in each state.

3. Inform Your Insurance Carriers

There is always a residual doubt about contacting insurance carriers too early in a dispute. If it all clears up without recourse to the courts—and a lot of problems do—was it wise to alert the company of your close call, causing them to regard you as a greater risk and possibly raise your premiums?

While it may not be the first call you make, be sure to inform your carrier in a timely fashion for your own safety. Insurance contracts are established on the basis of "*uberrimae fidei*," or "in utmost good faith," which means that if you don't keep your agent fully informed of your legal escapades, they may have grounds to vacate the policy and leave you without coverage protection.

4. When You Need Information

When problems occur, there is a tendency to try to find out as much as possible about the issue in question. While legal advice is going to be high on the list, there are other forms of information which may prove useful as well. Does your insurance carrier provide an information service? Some universities or professional organizations likewise may be able to help. It is unlikely that they will give you specific advice, but they may be able to point to individuals or sources that could be valuable for background purposes. Similarly, consumer advice agencies, public agencies (such as building inspection) and libraries may all be useful in certain circumstances in providing expert, and often free, information.

5. When You Need a Lawyer

At some point, it may be advisable to engage legal experts to advise you on how to proceed. In addition to selecting your counsel well, prepare yourself before the first meeting. Try to organize the main issues in dispute in a clear narrative, preferably in writing. This should be accompanied by a chronological account of the salient events that led to the dispute. While attorneys tend to be quick learners, the complexity of a construction dispute may take a while to explain, and sorting out the issues in your own mind before you start saves both time and money.

Take any important documents with you to the meeting in an accessible format (like a file folder or disk), chronologically arranged and clearly labeled. If there are a lot of documents, an index will help the attorney to work through the details quickly and efficiently.

Finally, work out before the meeting any questions or points of clarification you want answered by your attorney and commit them to writing. Otherwise, in the heat of discussion, you may forget to raise them.

6. Consider the Options

While the subpoena sitting on the desk seems to indicate that a courtroom appearance is in your future, there is plenty of time to explore alternative methods of dispute resolution. Arbitration is an attractive option that might be possible, and mediation should also be explored as an alternative to litigation. Settlement is another alternative, of course, and needs to be discussed fully before any formal action is taken.

In any event, react as calmly as possible should legal action loom, use the many information sources and options available to you and, in a deliberate, calm manner, select an approach to dealing with the issues that is rational and the least wasteful in time, money and nervous energy.

Question & Answer

There is so much information around these days on the Web, in the mail and on my desk. Surely I don't have to know all of it?

Despite a plethora of information available on the Web and generated through new technological achievements, the architect is expected to have an extensive and up-to-date knowledge of data relevant to the design and construction processes (see page 10).

In the early design phases, the architect's role of advising clients also implicitly involves making sure that clients are aware of their responsibilities too. This means that, while the AIA contract specifies that the client is responsible for the program, site surveys, easements, etc., the architect should ensure that the information necessary for a successful project is available. Are the client's needs fully expressed in the program? Are the site surveys up to date? Will any easements be necessary to construct the project, or are any restrictive covenants in place likely to affect the design? Can the decisions of the consultants be relied upon (because if *you* hired them, you are vicariously responsible for their actions)?

While architects are not necessarily expected to know everything, it is reasonable to expect that they will know what they *don't* know and therefore ask some questions, do some research, hire appropriate experts, or at least alert others to the need to find out more facts.

Completion

7

COMPLETION

When the contractor is nearing completion of the work, a number of procedures are recommended to ensure smooth completion of contractual performance. These procedures are the same for full completion, and partial completion where a designated portion of the work may be ready for occupation.

Substantial Completion

As soon as the contractor decides that the project has reached a state of substantial completion (i.e., when the owner can occupy or utilize the work for its intended purpose), a punch list is prepared. The punch list contains details of all outstanding items that the contractor intends to complete or correct, and it is sent to the architect who may then make arrangements for inspection.

The architect can amend the punch list and add extra items, if necessary. In the event that the architect feels that the work is not substantially complete, the contractor will be informed, and the architect need not return to reinspect until sufficient evidence is available to suggest that the work has reached the required standard. When the architect's inspection indicates that substantial completion has been reached, the Certificate of Substantial Completion will be issued (AIA Document G704, see page 108). This is prepared by the architect and sent to the owner and the contractor. The Certificate of Substantial Completion is an important document which has an effect upon:

- The contractor's warranty period
- The architect's liability period (in some instances: see page 109)
- The responsibilities of the owner and the contractor in respect of site security, insurance, heat and utilities, damage to the work, and maintenance

The architect should take care when issuing the certificate to ensure that the work has in fact been substantially completed. The certificate will establish the date of substantial completion and indicate the time allowed to complete the outstanding work. Following receipt of the certificate, the contractor can apply for payment in the normal way (see page 96). This payment should take account of the retainage agreed upon in the contract documents.

Final Completion

When the items on the punch list have been completed, the contractor should notify the architect in writing. The architect must then promptly inspect the work and, if it appears to be in conformance with the contract drawings and specifications, will issue a final certificate for payment, which is usually in the form of a letter. Upon issuance of the final certificate, the contractor becomes entitled to payment for all outstanding sums. However, certain states' lien laws may make it desirable to withhold a percentage of the retainage for a period of time. If this is considered necessary, it should be stated in the bidding documents and in the owner-contractor agreement.

Before final payment is made, the owner and architect should carefully check that:

- all required certificates of inspection, bonds, record drawings, and warranties have been delivered to the owner;
- keying schedule has been delivered (if not already undertaken);
- any instructions regarding operation of equipment have been supplied;
- all accounts have been adjusted (contract sums, deductions, change orders, deductions for uncorrected work: AIA Document A201, 9.10).

Before the final payment is made, it is also usual to take certain safety measures. These include:

- ensuring that the owner is protected from all possible lien claims (AIA Document G706A: see page 105);
- requiring an affidavit that all wages, and bills for materials and equipment (or other debts connected with the work which might conceivably revert to the owner) are paid in full.

If any payments by the contractor are still outstanding, the owner may require indemnification against third party claims. As an added precaution, the consent of the contractor's surety should be obtained prior to final payment (AIA Document G707a, Consent of Surety to Reduction in or Partial Release of Retainage, may be used).

Additional safety measures may be taken by the owner depending upon:

- The contractor's reputation
- Customs and practices of the area
- State lien laws
- Owner requirements

If there are any exceptions to the normal procedures recommended by the AIA relating to final safeguards, a payment or bond by the contractor can be used to discharge further responsibility.

The 1997 edition of B141 also provides for two meetings between the architect and the owner at this stage. One is held after Substantial Completion to review the need for facility operation services, and the other takes place within one year of Substantial Completion to review the building's performance.

Final Payment

Final payment by the owner of the balance of the contract sum, plus any remaining retainage, constitutes a waiver of all claims against the contractor except for:

- Unsettled liens
- Faulty or defective work appearing after substantial completion
- Failure of compliance with the contract documents
- The terms of any special warranties that may have been provided

Similarly, acceptance of the final payment by the contractor waives all rights to any further claims against the owner, with the exception of any claims made in writing at the time of the final application for payment. At the completion of the construction work, the architect submits a final account to the owner for outstanding payments. Any work undertaken beyond this time forms the basis for additional compensation, and may include:

- Furnishing a set of amended "as-built" drawings for the owner's records, which may be useful if further work or adaptation are anticipated.
- Site visits and advice to the owner concerning work to be undertaken by the contractor during the 12-month warranty period
- Inspection of the project prior to the expiration of the warranty period, and compilation of a report listing necessary repairs or corrections. An inspection may be arranged by the architect to check the work when complete (see A201.2.7).
- Maintenance advice or reports may be undertaken, possibly on the basis of an annual retainer.
- Post-occupancy evaluations may be carried out once the building is in use to judge its success and general performance.

At the completion of the project, the contractual relationship between the architect and the owner comes to an end. However, a period of continuing liability follows under the law of tort, during which time the architect may still be liable for negligent acts or omissions. The length of this continuing liability period is established by individual state law, and may vary considerably (see page 109).

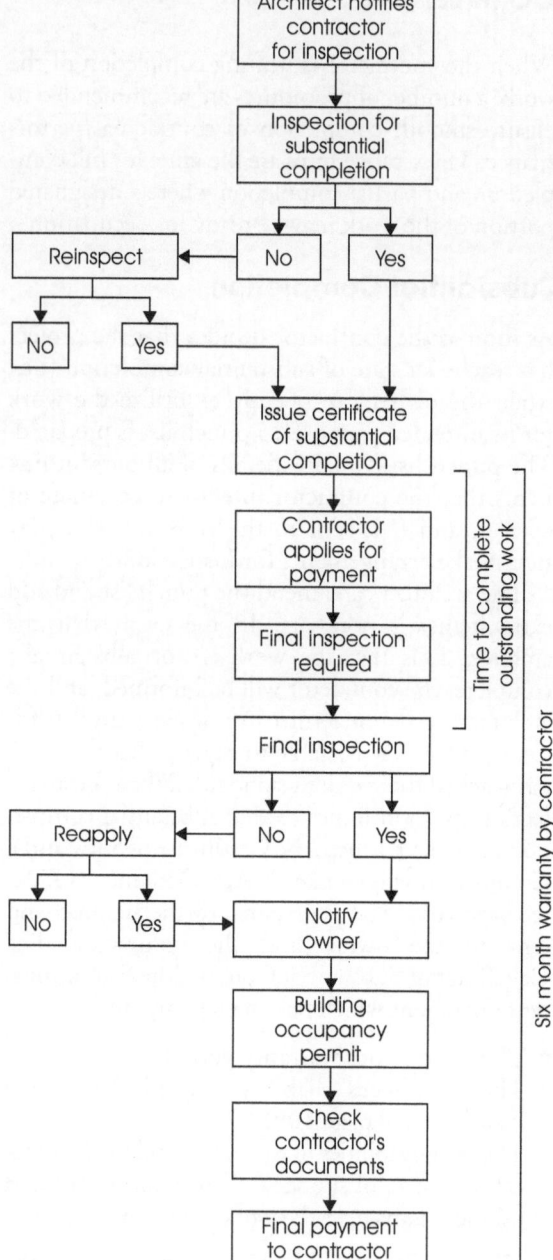

Figure 7.1

The American Institute of Architects is pleased to provide this sample copy of an AIA Contract Document for educational purposes. Created with the consensus of contractors, attorneys, architects and engineers, the AIA Contract Documents represent over 110 years of legal precedent.

CONTRACTOR'S AFFIDAVIT OF PAYMENT OF DEBTS AND CLAIMS

AIA Document G706

(Instructions on reverse side)

OWNER	☐
ARCHITECT	☐
CONTRACTOR	☐
SURETY	☐
OTHER	☐

TO OWNER:
(Name and address)

ARCHITECT'S PROJECT NO.:

CONTRACT FOR:

PROJECT:
(Name and address)

CONTRACT DATED:

STATE OF:
COUNTY OF:

The undersigned hereby certifies that, except as listed below, payment has been made in full and all obligations have otherwise been satisfied for all materials and equipment furnished, for all work, labor, and services performed, and for all known indebtedness and claims against the Contractor for damages arising in any manner in connection with the performance of the Contract referenced above for which the Owner or Owner's property might in any way be held responsible or encumbered.

EXCEPTIONS:

SUPPORTING DOCUMENTS ATTACHED HERETO:

1. Consent of Surety to Final Payment. Whenever Surety is involved, Consent of Surety is required. AIA Document G707, Consent of Surety, may be used for this purpose.

    Indicate attachment: ☐ yes ☐ no

The following supporting documents should be attached hereto if required by the Owner:

1. Contractor's Release or Waiver of Liens, conditional upon receipt of final payment.

2. Separate Releases or Waivers of Liens from Subcontractors and material and equipment suppliers, to the extent required by the Owner, accompanied by a list thereof.

3. Contractor's Affidavit of Release of Liens (AIA Document G706A).

CONTRACTOR:
(Name and address)

BY: _____
(Signature of authorized representative)

_____
(Printed name and title)

Subscribed and sworn to before me on this date:

Notary Public:

My Commission Expires:

AIA CAUTION: You should sign an original AIA document that has this caution printed in red. An original assures that changes will not be obscured as may occur when documents are reproduced. See Instruction Sheet for Limited License for Reproduction of this document.

Reproduced with permission of The American Institute of Architects, 1735 New York Avenue NW, Washington D.C. 20006. For more information or to purchase AIA Contract Documents, visit www.aia.org.

AIA Document G706: Contractor's Affidavit of Payment of Debts and Claims

▓AIA® Document G706A™ – 1994

Contractor's Affidavit of Release of Liens

PROJECT: *(Name and address)*

ARCHITECT'S PROJECT NUMBER:

CONTRACT FOR:

OWNER ☐

ARCHITECT ☐

CONTRACTOR ☐

SURETY ☐

OTHER ☐

TO OWNER: *(Name and address)*

CONTRACT DATED:

STATE OF:

COUNTY OF:

The undersigned hereby certifies that to the best of the undersigned's knowledge, information and belief, except as listed below, the Releases or Waivers of Lien attached hereto include the Contractor, all Subcontractors, all suppliers of materials and equipment, and all performers of Work, labor or services who have or may have liens or encumbrances or the right to assert liens or encumbrances against any property of the Owner arising in any manner out of the performance of the Contract referenced above.

EXCEPTIONS:

SUPPORTING DOCUMENTS ATTACHED HERETO:

1. Contractor's Release or Waiver of Liens, conditional upon receipt of final payment.

2. Separate Releases or Waivers of Liens from Subcontractors and material and equipment suppliers, to the extent required by the Owner, accompanied by a list thereof.

CONTRACTOR: *(Name and address)*

BY:

_____
(Signature of authorized representative)

_____
(Printed name and title)

Subscribed and sworn to before me on this date:

Notary Public:

My Commission Expires:

CAUTION: You should sign an original AIA Contract Document, on which this text appears in RED. An original assures that changes will not be obscured.

Document G707™ – 1994

Consent of Surety to Final Payment

PROJECT: *(Name and address)*   ARCHITECT'S PROJECT NUMBER:

OWNER ☐

CONTRACT FOR:

ARCHITECT ☐

CONTRACTOR ☐

TO OWNER: *(Name and address)*   CONTRACT DATED:

SURETY ☐

OTHER® ☐

In accordance with the provisions of the Contract between the Owner and the Contractor as indicated above, the
(Insert name and address of Surety)

, SURETY,

on bond of
(Insert name and address of Contractor)

, CONTRACTOR,

hereby approves of the final payment to the Contractor, and agrees that final payment to the Contractor shall not
relieve the Surety of any of its obligations to
(Insert name and address of Owner)

, OWNER,

as set forth in said Surety's bond.

IN WITNESS WHEREOF, the Surety has hereunto set its hand on this date:
(Insert in writing the month followed by the numeric date and year.)

_____
(Surety)

_____
(Signature of authorized representative)

Attest: _____
(Seal):   *(Printed name and title)*

**CAUTION: You should sign an original AIA Contract Document, on which this text appears in RED. An original assures that
changes will not be obscured.**

The American Institute of Architects is pleased to provide this sample copy of an AIA Contract Document for educational purposes. Created with the consensus of contractors, attorneys, architects and engineers, the AIA Contract Documents represent over 110 years of legal precedent.

2000 EDITION

AIA DOCUMENT | G704-2000

Certificate of Substantial Completion
(Instructions on reverse side)

PROJECT:
(Name and address)

PROJECT NUMBER:

CONTRACT FOR:

CONTRACT DATE:

TO OWNER:
(Name and address)

TO CONTRACTOR:
(Name and address)

OWNER ☐
ARCHITECT ☐
CONTRACTOR ☐
FIELD ☐
OTHER ☐

PROJECT OR PORTION OF THE PROJECT DESIGNATED FOR PARTIAL OCCUPANCY OR USE SHALL INCLUDE:

The Work performed under this Contract has been reviewed and found, to the Architect's best knowledge, information and belief, to be substantially complete. Substantial Completion is the stage in the progress of the Work when the Work or designated portion is sufficiently complete in accordance with the Contract Documents so that the Owner can occupy or utilize the Work for its intended use. The date of Substantial Completion of the Project or portion designated above is the date of issuance established by this Certificate, which is also the date of commencement of applicable warranties required by the Contract Documents, except as stated below:

ARCHITECT _____ **BY** _____ **DATE OF ISSUANCE** _____

A list of items to be completed or corrected is attached hereto. The failure to include any items on such list does not alter the responsibility of the Contractor to complete all Work in accordance with the Contract Documents. Unless otherwise agreed to in writing, the date of commencement of warranties for items on the attached list will be the date of issuance of the final Certificate of Payment or the date of final payment.

Cost estimate of Work that is incomplete or defective:

The Contractor will complete or correct the Work on the list of items attached hereto within
_____ (_____) days from the above date of Substantial Completion.

CONTRACTOR _____ **BY** _____ **DATE** _____

The Owner accepts the Work or designated portion as substantially complete and will assume full possession at _____ *(time)* on _____ *(date)*.

OWNER _____ **BY** _____ **DATE** _____

The responsibilities of the Owner and Contractor for security, maintenance, heat, utilities, damage to the Work and insurance shall be as follows:
(Note: Owner's and Contractor's legal and insurance counsel should determine and review insurance requirements and coverage.)

©2000 AIA®
AIA DOCUMENT G704-2000
CERTIFICATE OF
SUBSTANTIAL COMPLETION

1

The American Institute
of Architects
1735 New York Avenue, N.W.
Washington, D.C. 20006-5292

AIA Document G704–2000: Certificate of Substantial Completion

PRACTICE OVERVIEW

LIMITATION OF LIABILITY

A survey of prominent cases involving design professionals over time indicates a broad range of areas within practice where litigation has arisen, some of which, a few years ago, would have appeared unlikely areas of vulnerability. However, the increased duty of care expected of design professionals to the level where "reasonable" behavior has been raised to new heights of expected performance, has given rise to claims in some quite unexpected areas. For example, negligent specification of materials has become an area of concern, while approval of shop drawings has also resulted in legal action. In the most extreme cases, there were reports of the engineer who inspected the roof of the Hyatt Regency, Kansas City, being enjoined in a law suit for the collapse of the skywalk despite no inspection on his part of the faulty structure; similarly a firm of architects in Philadelphia were sued for designing a shopping mall "conducive to kidnapping" following a crime committed there.[1]

Beyond these incidents, however, there is one area of concern which, when viewed in context, has increased in prominence and magnitude. In a study undertaken in Wisconsin over a twenty-year period, it transpired that 23 percent of the cases involving architects taken to the Supreme Court concerned the question of limitation of liability. Similarly, a review of other cases nationally indicates a considerable degree of debate and activity within this area.

A limitation period is expressed in statute form in each state, after the expiration of which no legal action can be brought by an aggrieved person. It was developed and expanded largely in the 1950s as the doctrine of privity of contract was eroded to give the design professions some degree of protection from indefinite legal threat. As the statutes are enacted on a state-by-state basis, they vary considerably both in the time periods they specify, and in the allocation of time for specific actions. For example, differing limitation periods may be specified for slander, bodily injury, property damage, etc. (although some actions may fall under two or more categories), and may vary in duration from two to fifteen years, depending upon the state in which they are enacted.

Some states enact *statutes of repose* as opposed to *statutes of limitation* specifically for work undertaken by the design professions. The central difference lies in commencement of action; for the latter, *an event* establishes the date from when the time period for action begins; in a statute of repose, however, *a specific date* will activate the period, irrespective of any fault or action on the part of the designer. This may, of course, mean that the limitation period could have expired before building failure or damage have occurred, thus depriving the aggrieved party of a remedy. The potential problems of constitutionality here, the deprivation of the plaintiff's "day in court" as opposed to the affording of some degree of legal protection to the design professional, has led to a number of challenges to these statutes, and a number of interpretations as to their meaning and purpose. Although these cases have been decided in various states, they collectively provide a broad picture of the variation in interpretation which presently exists in relation to when the statutory period begins. If AIA contract documentation is used, provisions are included to establish the date at the end of the construction of the work upon certification of substantial completion, although in some cases legal action against the architect has been allowed where the limitation period was conceived as beginning at the end of the professional relationship, which could include a period where the architect gives the client post-completion advice on problems arising from the construction.[2] Architects providing continuing services on a number of projects are particularly ill served by this ruling. Furthermore, some courts have accepted the Injury Rule as being applicable, establishing the construction of the limitation period as commensurate with the actual failure of the building, as would be the case in any third party injury case.[3]

However, perhaps the most worrying development in this field for design professionals concerns the "Discovery Rule." Here, the date of commencement is established at the time the plaintiff discovers, or should have discovered, the fault. The rationale for the rule lies in the complexity of building construction, the potential difficulties involved in determining faults which may be covered over or be underground and the potential time lag from completion to discovery. In such cases, by the time the plaintiff has realized the impact of the fault in the absence of the Discovery Rule, he or she may be statutorily deprived of a remedy. The Discovery Rule allows for both fault and damage to be taken into consideration and has been used successfully in a number of states.

The implications of the Discovery Rule have had a dramatic impact upon architectural practice. Buildings completed years previously may suddenly develop signs of failure;[4] from this point, the client will have a statutorily set number of years in which to make a claim, thereby providing the architect with the prospect of virtually unlimited future liability. In the past, architects who insured against potential legal suits could allow coverage to lapse at the expiration of the statutory period related to each project; retiring professionals also could terminate the coverage after this period, confident of no further claims against them. With the emergence of the Discovery Rule, however, such time-related protection no longer exists, and architects may be faced with potential claims long after their retirement. This point is most forcibly brought home in an English case, where an architect was sued eleven years after his death, and his estate, which supported his wife in old age, was threatened.[5]

There have, fortunately, been some developments in the past few years that have given some relief to the situation. Analyses of the problem of longevity of exposure have yielded some comforting findings. For example, the majority of claims (95 percent) are typically raised in the first ten years following completion. Beyond that, natural deterioration, poor maintenance and a hazier recollection of events by witnesses and relevant parties make building a convincing case less easy. Consequently, the idea of a long-stop statute has been successfully initiated in many states and other countries.

In Scotland, for example, the Prescription of Limitation (Scotland) Act 1973 provides a statutory period of twenty years for claims to be made. In the United States, some jurisdictions have extended the period during which architects may be held accountable (in Wisconsin, for example, from six to ten years), but have specified that it begins upon the date of substantial completion. This provides an extended period of time in which a suit may be initiated, but provides the architect with a guaranteed date of limitation expiration. Statutes of repose have still been legally challenged, but provide a measure of predictability to architects in dealing with their liability exposure.

References

1. *Wall Street Journal*, 6 December 1983, 21.
2. *County of Milwaukee* v. *Schmidt Garden and Erikson* 4 3 Wi 2d 445, 168 NW 2d SS9 (1969).
3. *Abramowski* v. *Wm. Kilps Sons Realty, Inc.* 80 W.S. 2d 468; 2S9 N.W 2d 306 (1977).
4. *Rosenberg* v. *Town of Bergen*, 61 NJ 190, 293 A2d 662 (1972).
5. Cecil, R., "Writing your Will to Defend your Estate from Eternal Liability," *Royal Institute of British Architects Journal*, December 1982.

Question & Answer

Seems like every other client I have is slow to pay bills on time. Are there any strategies I could use to speed things up and get paid more regularly?

Getting paid can be a headache for architects. Almost a third of legal cases involving architects concern fee collection and estimates have run as high as $70 million a year for uncollected fees.

In some ways, it's not surprising that this is such a problem. Many projects do not pass beyond the design phase—too expensive, insufficient financing, changes of plan, etc.—and clients might be reticent to pay for ideas which they will now never use.

Hopefully, a good contract and regular payment schedule can minimize problems but if all else fails, some states allow for a mechanic's lien to be placed on the owner's property to force payment. Of course, if no construction has begun it may not be enforceable, so the architect may have to consider other strategies to coax payment from a recalcitrant client:

A Firm Request

A polite but increasingly firm series of letters that request payment, outlining the services provided and a schedule of necessary payment can lay the foundations for a successful claim if it ultimately fails. The tone of the letters should be professional, and remember that you are laying a "paper trail" that will hold up in court in later years should your blandishments fail.

Dispute resolution

The courts are always an option, although time and expense make them a last resort rather than an opening strategy. Remember that alternatives exist in both arbitration and mediation as possible means of dispute resolution, but be reassured that studies show that, when an architect sues for fees, the likelihood of success, if decided cases are anything to go by, has been as high as 75 percent.

Collection Agencies

Many architects avoid the option of collection agencies, believing them to be incompatible with a professional service. They can be effective, however, although expect to pay a significant percentage of the amount to the agency.

Forget it

If a client refuses to pay, despite every effort to settle on your part, you will at some point have to decide—do we sue or not?

Legal action is expensive (even arbitration and mediation have their costs), time-consuming and potentially damaging to a reputation, and some practices choose to "eat" the loss rather than pursue a client in the courts.

While the least palatable choice for architects unjustly denied their fees, it is still a realistic option that must be considered.

Dispute resolution

Dispute resolution

TERMINATION

For a number of reasons, not all building contracts are fully performed as intended. A contract can be terminated in a variety of ways, e.g., by agreement (see page 63), but the AIA General Conditions make special provision for the unilateral termination of the contract by either the owner or the contractor in the event of specified circumstances.

Termination by the Owner (AIA Document A201, Article 14.2.1)

The owner may be permitted to terminate the contract:

- if the contractor is adjudged bankrupt;
- if the contractor makes a general assignment for the benefit of his or her creditors;
- if a receiver is appointed on account of the insolvency;
- if the contractor persistently or repeatedly fails to supply properly skilled workers or proper materials (unless an extension has been granted);
- if the contractor fails to make prompt payment to subcontractors and/or suppliers;
- if the contractor persistently disregards laws, rules, ordinances, regulations, or orders of a public authority;
- if the contractor is guilty of a substantial violation of one of the provisions of the contract documents.

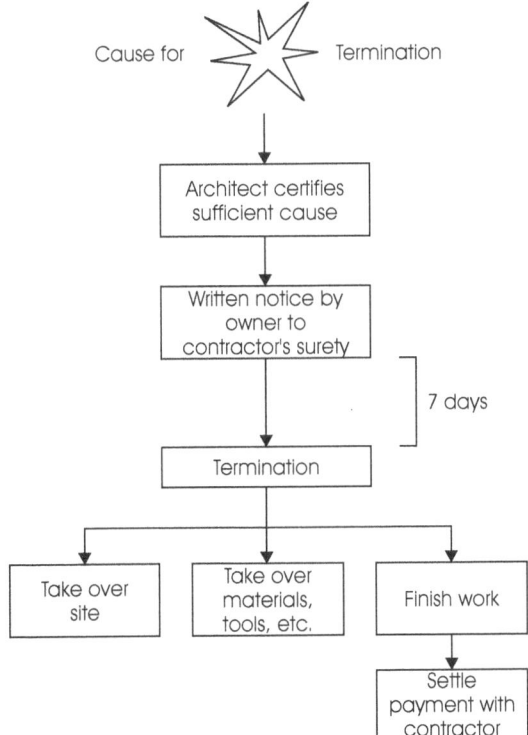

Figure 8.1

Procedure

Under the AIA General Conditions, the owner must seek from the architect certification that sufficient cause exists to justify termination of the contract. It should be noted that the United States Bankruptcy Code provides that trustees in bankruptcy may assume or assign contracts provided that any pre-bankruptcy defects have been cured. As a result of this change in the law, the AIA has advised that Article 14.2.1 has been effectively invalidated insofar as it relates to bankruptcy. Termination by the owner, therefore, should only be undertaken with the assistance of legal counsel, who should carefully review the circumstances in the light of the contract itself, and relevant state and federal law.

If the decision is made to terminate, seven days after written notice has been sent to the contractor and any surety, the owner may:

- Terminate the contract with the contractor
- Take possession of the site
- Take possession of all materials, equipment, tools, construction equipment, and machinery on the site owned by the contractor
- Finish the work in the most expedient way

The contractor will not be entitled to any further payment of outstanding fees until the project is complete.

Costs

If the monies owing to the contractor exceed the cost of finishing the work (including additional fees of the architect), the contractor will be reimbursed the difference. However, if the cost of finishing is higher than the amount owed to the contractor, the contractor will be liable to pay the excess, the sum of which must be certified by the architect.

Termination by the Contractor (AIA Document A201, Article 14.1.1)

The contractor may also have the right to terminate the contract under the following circumstances:

If the work is stopped for a period of thirty days due to:
- Order of the courts
- Public authority intervention
- The result of an act of government (e.g., declaration of national emergency, making material unavailable)
- Through no act of the contractor or his or her subcontractors or agents

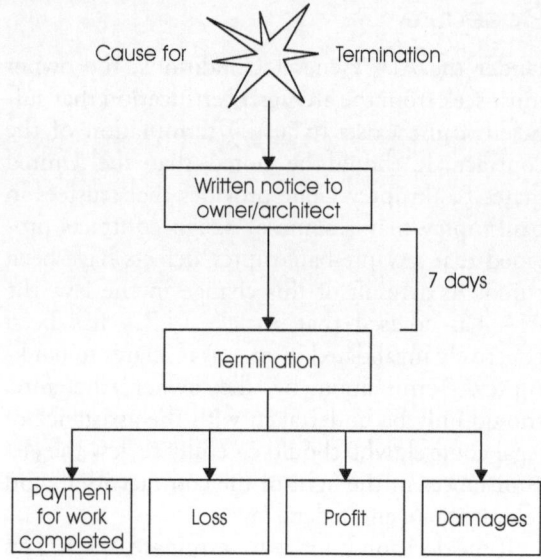

Figure 8.2

Or, if the work is stopped for thirty days because:

- The architect has not issued a certificate for payment
- The owner had not paid the amount certified

Procedure

After seven days' written notice to the owner and the architect, the contractor may terminate the contract and recover from the owner:

- Payment for all work executed to date
- Proven loss sustained in the expenditure for materials, equipment, tools, construction equipment, and machinery
- Reasonable profit
- Damages

Termination is a drastic step to take in the event of contractual disputes, and should be given extremely careful consideration. The aggrieved party should ensure that all procedures required by the contract documents and by relevant laws are strictly adhered to in order to prevent successful counterclaims.

DISPUTE RESOLUTION

If all other contractual mechanisms fail to provide satisfactory resolution of a dispute between the contracting parties, the introduction of a third party may be necessary to settle the matter. In addition to litigation, both arbitration and mediation can be effective in resolving disputes.

ARBITRATION

The third party could be a civil/court judge, if the normal court procedures are followed (see page 4).

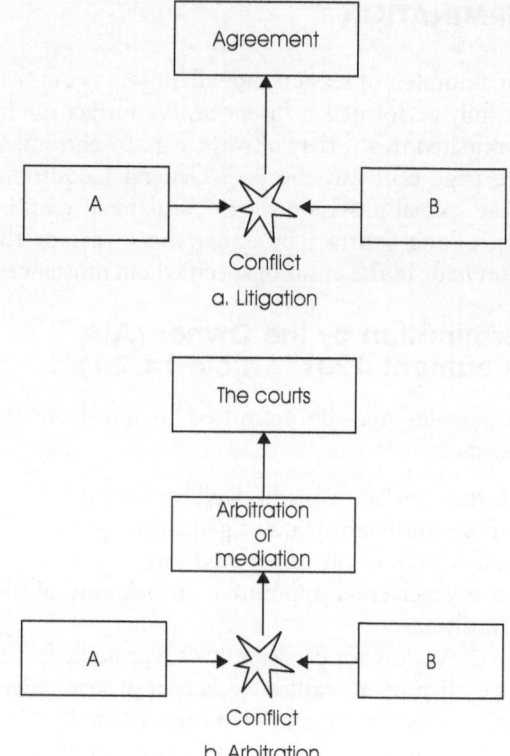

Figure 8.3

Alternatively, a dispute could be settled by the relatively less formal procedures of arbitration or mediation.

AIA Document B141 (1.3.4.1) states that disputes must first be attempted through mediation before resorting to arbitration or other legal proceedings.

Whereas the courts form part of the United States' judicial system and are, therefore, subject to all of its procedural and administrative rules, disputes submitted to arbitration can be settled by an informal private hearing in the presence of whoever the parties choose. If agreement is not possible, a designated third party may select the arbitrator.

The arbitrator is usually someone with specific knowledge and experience in the field in which the dispute has arisen. In the building sector, this might be an architect, engineer, or other professional person who is usually a member of the American Arbitration Association, a national organization which operates a commercial panel that serves, among other things, the building industry.

Advantages of Arbitration

The major advantages of submitting a dispute to arbitration are:

- Privacy
- Convenience

- Speed
- Expense
- Informality
- Expertise

Privacy

Trade secrets and reputations may be shielded from the public in a private arbitration. The courts, however, are public forums and privacy is generally not possible.

Convenience

Arbitration hearings can be held anywhere to suit the parties, such as at the site of the dispute.

Speed

Disputes can be handled quickly, without the inconvenience of having to fit into a court's schedule. In projects where time is of the utmost importance, this can be a decisive factor.

Expense

Money might be saved in two ways:

1. The potentially lower cost of the hearing
2. The speedy resolution of the dispute

Informality

Courtroom procedures may be dispensed with or modified at the direction of the arbitrator.

Expertise

Difficult construction-oriented problems may be more readily understood by an arbitrator experienced in the construction field than by a professional judge.

Disadvantages of Arbitration

The disadvantages of arbitration are:

- Cost
- Lack of legal expertise
- No binding precedent

Cost

Aside from the expense of legal counsel, expert witnesses, etc., the arbitrator's fees must be paid together with the cost of hiring the place of the hearing. In the court system, the services of the judge and the use of the courtroom are not additional expenses.

Lack of Legal Expertise

Though knowledgeable in the field of the dispute, the arbitrator may be less well informed with regard to the law than a professional judge.

No Binding Precedent

Each case submitted to arbitration is decided upon its own merits, without necessarily any regard to previous cases. This can make it difficult for the parties to ascertain the strength of their arguments.

When to Arbitrate

Parties may go to arbitration:

- After attempting to resolve the dispute through mediation (B141, Article 1.3.4.1)
- By agreement after the dispute has arisen
- By agreement before the dispute arises (i.e., as a condition of the contract)
- By order of court (many states will enforce an agreement to arbitrate)

Agreement to arbitrate prior to a dispute occurring is the preferable method, and most building contracts provide for arbitration proceedings by stating that the parties agree to be bound by the decision of an arbitrator in the event of disagreement (AIA Document A201, Article 7.9.1). Many AIA standard forms of contract provide for arbitration, including the owner-architect agreements.

In addition to the agreement to arbitrate, both parties to AIA construction contracts agree to abide by the Construction Industry Arbitration Rules which are published by the American Arbitration Association. However, some state laws regarding arbitration vary, and this should be taken into account at the contract formation stage in case any modifications may be necessary to match state requirements. The assistance of legal counsel is advisable.

Arbitration Procedure

Either party to the AIA construction contract (not necessarily with the consent of the other party at the time of the dispute) may initiate arbitration proceedings by writing to the other party, with a copy to the architect, within the time allowed by the contract documents (AIA Document A201, Article 2.2.12). This letter usually includes:

- The reason for the dispute
- The amount involved
- The remedy sought

It represents a Notice of Demand for Arbitration under the terms of the AIA contract.

Two copies of the notice should be filed with the American Arbitration Association (AAA) within seven days of the notice, sending a copy of the answer to the originator of the proceedings. If the respondent chooses not to reply, there is nonetheless an assumption that the claim is denied.

Although arbitration proceedings are in progress, both parties are constrained by the contractor to meet all their contractual obligations unless otherwise agreed in writing, or unless the reason for the arbitration is the breakdown of the contract itself.

Selection of the Arbitrator

An arbitrator may be selected:

- By agreement of the parties before the dispute
- By agreement of the parties during the dispute
- By reference to the American Arbitration Association

In the latter case, the AAA sends a list of possible arbitrators to both parties who are given seven days to delete any names they consider to be unacceptable, and list the remaining names in order of preference. The AAA then contacts an arbitrator (or arbitrators: a panel of three is sometimes selected) on the basis of the amended lists. In the event that none of the names are acceptable, the AAA will appoint an arbitrator without submitting new lists.

Prior to accepting the appointment, prospective arbitrators should assess their suitability for the case, and disclose all potential conflicts of interest (e.g., personal knowledge of one of the parties, or a financial interest in the dispute).

Pre-Hearing Procedures

A pre-hearing conference may be arranged at the parties' request, or if the AAA believes such a conference would be useful. The pre-hearing conference allows for an exchange of information, the stipulation of uncontested facts, and the agreement of administrative details such as:

- Locale: This may be mutually agreed upon, but in the event of disagreement between the parties, the AAA will make binding decision.
- Use of legal counsel: This is acceptable in many states, but if one party decides to engage a legal representative, the other party and the AAA must be notified at least three days prior to the hearing.
- Stenographic record: If one of the parties requests a record of the proceedings, the requesting party must bear the costs unless both parties agree to share the expense.

- Time and place: The arbitrator decides the time and place of the hearing, and the AAA will notify the parties at least five days in advance.

The Hearing

The hearing should only be held when all the requisite documents have been exchanged. In the event of the refusal of one party to participate, or if there is an attempt to deliberately obstruct the proceedings, the arbitration may continue *ex parte* (i.e., on the proof of one party only) provided that the absent party has been notified in writing of his or her right to attend. An award may not be made simply on the basis of the absent party's default, and all relevant evidence should be heard by the arbitrator prior to making the award.

The hearing generally comprises the following stages:

- The oath of the arbitrator (if required by the parties)
- Recording of time, place, date of hearing, the parties present and statements of claim and response
- The arbitrator may ask for statements from both sides outlining the issues involved in the dispute
- The claimant will then present the claim, supported by proof in the form of testimony, exhibits, etc.
- The claimant's witnesses will be examined, cross-examined (by the respondent or counsel) and then re-examined by the claimant
- The respondent must then follow the same procedure for the defense and counterclaim (if any).

Inspection of property may be required, and both parties are generally given the opportunity to accompany the arbitrator. If no other proof is required or forthcoming, the arbitrator will close the proceedings and make a decision within the specified time (usually not later than thirty days after the hearing). No communication between the arbitrator and the parties to the dispute should take place, except through AAA.

The Award

The award should be made in accordance with relevant state law and will be sent to both parties simultaneously by the AAA. Typically, the parties will be asked to deposit a sum with the AAA at the beginning of the arbitration proceedings to ensure payment of the arbitrator.

In the event that one of the parties refuses to accept the arbitrator's decision, application may be made to the courts to enforce the award.

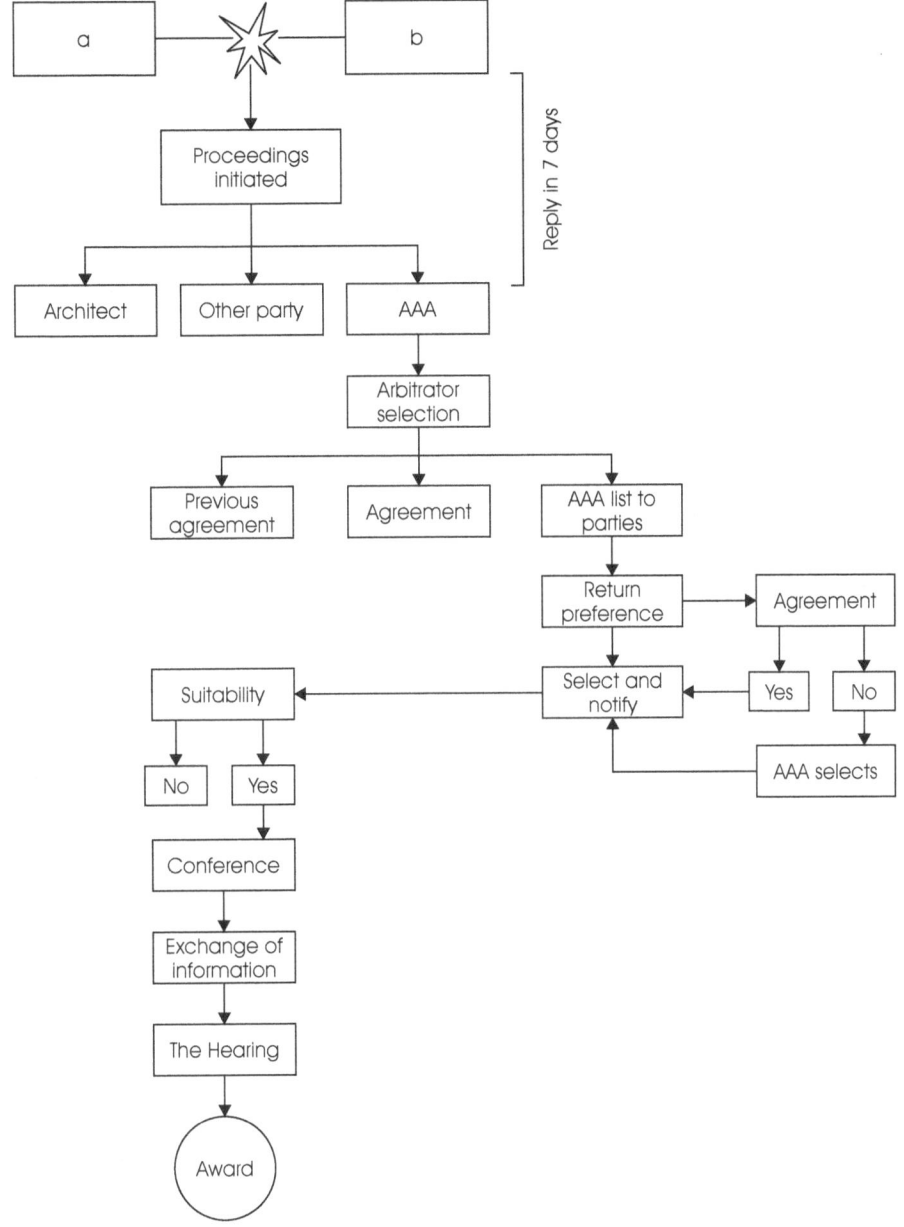

Figure 8.4

Although arbitrations are carried out largely independently of the court system, the courts may have statutory power to reject or vacate the arbitrator's award in the following circumstances:

- If the arbitrator exceeds his or her authority
- If there is evidence of corruption, fraud, or partiality
- If the arbitrator refuses to hear evidence of either party
- If the arbitration agreement is improper

Modifications to the award may be allowed by the arbitrator if a party considers that a mistake has been made. However, awards are usually reaffirmed.

MEDIATION

While dispute resolution has tended to focus on litigation and arbitration and their relative advantages, a new alternative is growing in use in the construction industry.

Despite their differences, litigation and arbitration use basically the same principle to resolve a dispute. Parties refer their differences to a third party who, after reviewing the evidence, usually declares one side a winner, the decision being subject to appeal under certain conditions.

In the mediation process, disputing parties will engage the services of a third party, but the mediator will have *no* authority to make a decision

---

American Arbitration Association, Administrator
Commercial Arbitration Tribunal

Wesmey-Shovelgon Construction Co.
and
Acme Estates Inc.
Case Number 4136-1321-88

AWARD OF ARBITRATOR

I, THE UNDERSIGNED ARBITRATOR, having been designated in accordance with the Arbitration Agreement entered into by the above-named parties, and dated May 14, 2003, and having been duly sworn and having heard the proofs and allegations of the parties, AWARD as follows:

1. Within fourteen (14) days from the date of transmittal of this Award to the parties, Acme Estates Inc. shall pay to Wesmey-Shovelgon Construction Co. the sum of FOURTEEN THOUSAND THREE HUNDRED AND FORTY-SIX DOLLARS ($14,346.00) plus interest thereon at the rate of eleven and one quarter per cent (11.25 per cent) per annum from the date when construction work was stopped by Acme Estates Inc., that being May 15, 2004, until August 21, 2004.

2. The counterclaim of Acme Estates Inc. against Wesmey-Shovelgon Construction Co. is hereby denied.

3. The administrative fees of the American Arbitration Association amounting to SEVEN HUNDRED AND FIFTY-FOUR DOLLARS AND THIRTY FIVE CENTS ($754.35) shall be borne entirely by Acme Estates Inc.

4. This award is in full and final settlement of all claims and counterclaims submitted to the arbitration.

Signed: _____

Arbitrator _____

Date: _____

Notarized: _____

Note: The execution of the award may vary according to the legal requirements of the state in which the arbitration takes place.

Figure 8.5

which is binding on the individuals involved. In fact, the mediator's role is to enable the parties to voluntarily explore settlement options and jointly craft a resolution that resolves the dispute. Mediators have no authority beyond their powers of persuasion and creative problem-solving, and the process relies on the will of the disputing parties to resolve their differences without resorting to the more formal options of arbitration and litigation.

Despite the lack of mediator authority, the process has proven to be very effective if parties agree to try it, yielding a success rate of between 85 percent and 95 percent.

Initiating Mediation

Parties can agree at any time to submit their differences to mediation, although some contracts, such as the AIA Standard Form of Contract, now include a mediation clause as a prerequisite to arbitration or litigation. If one of the parties to such a signed contract tries to go directly to court without attempting mediation, the courts will likely delay legal action until a mediation has taken place.

The Benefits of Mediation

The primary advantages are:

- Disputing parties can create their own settlement.
- The mediator is a neutral third party, and can therefore help both parties explore alternative solutions.
- Mediation is very informal and can be arranged quickly.
- Costs can be dramatically reduced in legal fees, time and expended energy.
- There is a greater possibility of maintaining the working relationship between the parties.
- The free-form nature of the process allows for creative solutions to be explored.

Selection of the Mediator

Mediators in the construction industry are usually experienced professionals—architects, engineers and attorneys—who are trained in dispute resolution. If they are part of a recognized organization specializing in resolving problems, such as the

American Arbitration Association, they will be required to undertake regular training and instruction to ensure continued competency and placement on the AAA's Construction Mediation Panel.

While parties choose anyone they wish to mediate their dispute—the process is entirely voluntary—if they use the services of a group like the American Arbitration Association, they will be sent the details of a qualified mediator who serves on the Construction Panel. Either party may object to the mediator until a suitable candidate, who both sides feel will be fair, impartial and effective, is found.

The Process

While the informality of the mediation process enables the discussions to follow any direction appropriate to the parties and their dispute, the process usually involves a conference where everyone comes together to attempt resolution. However, there may be some sharing of documentation and outlining of the relative positions (and possibly desired outcomes) ahead of time to allow the mediator to become familiar with the details of the case.

At the conference, the mediator outlines the procedures and basic ground rules involving:

- Presentation of each party's case
- Order of presentation
- Decorum
- Discussion of unresolved issues
- Use of caucuses
- Confidentiality

Each party is given an opportunity to present their case to enable the mediator to gather as many facts as possible. Following any decision between the parties and the mediator, the groups will often move to separate rooms, and the mediator will caucus with each party separately, shuttling back and forth between them and potentially bringing them back together again, searching for a solution.

During the caucuses, the mediator's role is to:

- Clarify each party's version of the facts

- Establish relative priorities and positions
- Question assumptions and loosen fixed stances
- Explore and possibly forward alternative solutions
- Seek trade-offs, face-saving strategies and win/win solutions
- Probe and challenge the validity of each position

The Settlement

If, with the assistance of the mediator, a workable resolution is reached, it is usually committed to writing and signed by all parties as soon as possible. The agreement may also be put in the form of a consent award if the American Arbitration Association is involved. The AAA will then make the necessary arrangements.

THE ARCHITECT AS ARBITRATOR OR MEDIATOR

As noted earlier, the architect has a quasi-arbitral role in the administration of the construction contract. In addition, the architect's professional qualification and experience in the construction field imply a knowledge and expertise which might provide the basis for arbitration or mediation work. The American Arbitration Association has regional offices throughout the United States which may be contacted by architects wishing to apply for training and inclusion on the Commercial Panel.

THE ARCHITECT AS EXPERT WITNESS

It is possible that an architect may be called as expert witness at an arbitration or mediation to give professional opinions regarding a building dispute. The expert witness is not usually personally involved in the dispute, and is paid for objective expertise and opinion which may be given in a written report or by oral testimony (see page 125).

PRACTICE OVERVIEW

SOLVING DISPUTES—IS THERE AN EASIER WAY?

Unfortunately, many architectural practices are no strangers to legal disputes. Construction is, after all, a messy, high-risk business with huge amounts of money involved. Problems, unanticipated outcomes and misunderstandings all too often end up in some form of dispute resolution. While the incidence of legal action has lessened in past years, there is still a noticeably high level of litigation involving architects, which should give the profession reason to look for more effective alternatives in resolving their disputes.

The civil court system provides the most traditional means of sorting out legal problems, but has some sobering consequences. Going to court can be breathtakingly expensive and, of course, painfully slow. In the months—sometimes years—that it takes to work through the process of taking depositions, waiting for court dates and enduring the legal proceedings, huge amounts of nervous energy and valuable work time can be consumed.

Arbitration is often touted as a viable alternative to litigation, and can be faster, cheaper and more convenient. It also has the advantages of privacy and possibly the expertise of an arbitrator familiar with construction procedures who can understand the complex, often technical, facts of the case. However, its detractors point out that the general lack of an appeals procedure and the possibility of getting an arbitrator ignorant of broader legal issues (this is a particular beef of attorneys, naturally) can make the outcome of the process uncertain, and that costs and delays can run about as high in arbitration as in a comparable legal case. Sadly, this can sometimes be the case.

However, in the experience of many arbitrators, the win/lose nature of the outcome of each case has been its most frustrating drawback. As the only powers that an arbitrator wields are the ability to deny or uphold a claim and order a monetary award, there is always a winner and a loser and, frankly, construction disputes are often a lot more complex than that.

As both litigation and arbitration ultimately involve the judicial determination of a dispute on a win/lose basis, it has been encouraging to see the development of a new field of dispute resolution taking effect in the construction industry. Mediation is a relatively new phenomenon, but one that has begun to catch on. It differs from the other two forms of dispute resolution in one important way—the mediator, unlike the judge or arbitrator, has *no* powers to make a judgment. His or her only role is to facilitate discussion, help the parties explore alternative resolutions to their problems and ultimately work with them to craft an agreement that is mutually acceptable. Ideally, they strive for a win/win situation. It's often not quite as rosy as that, but at least avoids the knock down, drag-out battle that leaves one side victorious, but both sides financially and emotionally bruised.

Mediation has been remarkably effective in construction-related disputes, yielding a 90 percent success rate when it has been employed. Its principal advantages are speed, flexibility and economy. If it works, parties are spared crippling legal fees and long periods of uncertainty and concern. They may even salvage their professional relationship, not just on the project in dispute but in future years, a phenomenon less likely after legal action.

Of course, not all disputes lend themselves to mediation—multiple party mediations are a particular challenge—and not all parties possess the attributes to work through the process. However, if parties are willing to agree to try to sit down and discuss the issues (the single biggest factor in the high success rate of mediation), and look for ways to resolve their differences, they have a strong chance of walking away at the end of the day—that day—with the dispute behind them. Of course, this will require an open-minded approach, a willingness (albeit grudgingly) to compromise and an ability to see the other party's point of view, however annoying and unpalatable. If all else fails, arbitration and the courts are still available, but at least the parties have tried to resolve their differences themselves, often with success. Typically, once parties have committed to the concept of mediation, the success rate is impressively high, even if the hearings take many hours to complete.

When a nasty dispute looms, architects should consider a serious discussion with their legal counsel as to the advisability of setting up a mediation before more formal action is taken. Here are the attributes needed to approach the process positively and productively.

1. *Be willing to sit down with "the other side" and talk about the dispute calmly and constructively*
   Easier said than done, if the dispute has already turned nasty. I have been involved in arbitrations and mediations when I have had to physically separate parties, request one of them to leave the room (once when it was in his own house!) or suspend the discussion until tempers calm down. Sometimes, it's too difficult for an aggrieved party to even be in the same room with the other side, let alone amicably discuss solutions with them. However, experience shows that if parties can initially agree to try to discuss the matter informally, that simple agreement can set them on the path to resolution.

2. *Be open-minded*
   Listen to the other side's perspective on the matter—things are rarely black-and-white in the complex world of construction—and try, for the sake of argument, to see their point of view. Maybe they have a point as well, and that may affect your perspective on the case.

3. *Forget revenge*
   However personal the dispute has become, you have to leave emotion at the door. If you want to see personal retribution or are focused on a point of principle, the mediation will fail. A dispute has to be seen as a tangled mess that the parties can unravel if they approach it professionally and dispassionately—leave personality out of it.

4. *Think out of the box*
   Mediation provides the parties with the freedom of choice—they can resolve their dispute any way they like and can look to less conventional ways to create a solution. Can a settlement be spread out over time in a series of payments? Can an ongoing professional relationship—the promise of future work—be maintained? I have even known a simple, sincere apology to be the lynch pin in a dispute. Again, a willingness to both propose and consider nontraditional ideas can really help the process. Remember, once the informality of mediation is abandoned for more formal methods, your fate is in the hands of a third party judge or arbitrator and you have lost control of the decision-making process.

5. *Be prepared to compromise*
   No one likes to lose, but sometimes settling for less at this stage, even if you're convinced you have an ironclad case, may save you money in the long run. Balance the merits of a quick solution against even the best-case scenario—a clear win (never a certainty in either courts or arbitral hearings, despite what you believe or your attorney tells you) tempered, of course, by considerable legal bills. Is a compromise worth it, not just for the money, but for the time, effort and continued worry of a lawsuit you have saved? In this way, a compromise can be viewed as a win too.

6. *Look beyond the conflict*

Sure, you aren't that fond of the other party (client, contractor, etc.) now, but the construction world is small and life is long. You may well have had a good previous working relationship and, this dispute aside, will work together in the future. Does it make sense to preserve the relationship? Do you foresee working for or with them again? Sorting out a long-term strategy can help in going into a mediation with a view that transcends the dispute in question.

7. *Don't sweat the small stuff*

Avoid pettiness in the discussions. Don't let a resolution of the dispute be held up over a minor sticking point (quibbling over who said what in a conversation, for example). Keep thinking Big Picture—if I settle today I am free of further costs or worry about this matter and can get back to the business of architecture.

Mediation may not be the universal panacea for all construction disputes. Parties are not usually hugging after a resolution, but at least they can now move on, and often rebuild their relationship once things have calmed down and before any further acrimony has been exchanged. And while the success of mediation lies predominately in the attitudes of the parties to take control of their own settlement, it is important to involve an experienced mediator to manage the discussions between the parties, keeping discussions going, suggesting alternatives and being the catalyst for a productive settlement. Mediations can be lively occasions, involving shuttle diplomacy by the mediator, caucus meetings with the attorneys and even one-on-one exchanges between the two protagonists. A skillful mediator will orchestrate the pace of settlement, keep tempers under control, focus on positive settlement strategies and ultimately help the parties forge an agreement that clearly and irrevocably ends the dissent between them. While the sword of justice wielded in litigation and more quietly in arbitration is just as effective in creating a solution, the outcome is no longer in the parties' own hands. Isn't it worth giving mediation a try first?

Question & Answer

I've been asked to serve as an expert witness on a construction case. Is this something architects can get involved in?

Some architects become involved in court cases or arbitrations as expert witnesses, particularly where a professional opinion is necessary to help a judge, arbitrator or jury to decide whether certain levels of expected performance have been achieved. The use of an expert witness is common in negligence cases, where the standard of care—that is, the level of ordinary and reasonable skill usually exercised in practice—has to be established through professional opinion.

Being an expert witness is not always an easy role. Your opinions, which are remunerated, may carry the appearance of the "hired gun" and credibility will be judged against your credentials and experience relevant to the field of the dispute.

An expert witness's role might include:

- Helping the lawyer who retained you to review the case and develop case strategies
- Making site inspections
- Presenting a written report of your finding
- Giving deposition testimony to opposing counsel
- Advising on the use of technical construction terms
- Preparing questions for use in cross-examination of the opponent
- Listening to the opposing side's experts and recommending questions to undermine their testimony

Expert witnesses can refer to their notes during hearings. They should avoid jargon or being overcomplicated in their responses. They should not be patronizing or boring to the judge, jury or arbitrator, and should not exaggerate or provide inconsistent answers. To be as effective as possible, expert witnesses must be as accurate and honest as possible to avoid speculation during testimony upon their impartiality.

Glossary of common legal terms

ab initio	from the beginning
bona fide	in good faith
caveat emptor	let the buyer beware
ejusdem generic	of the same type
estoppel	a rule of evidence which prevents a person from denying or asserting a fact owing to a previous act
ex parte	upon the application of
ignorantia juris non excusat	ignorance of the law is no excuse
in personam	against a person, i.e., not against everyone
in rem	against a thing, i.e., applicable to everyone
inter se	among themselves
obiter dicta	things said by the way
per se	by itself
prima facie	on first view
quantum meruit	as much as he deserves
ratio decidendi	reason for the decision
res ipsa loquitur	the thing speaks for itself
stare decisis	to stand by past decisions
sui juris	of legal capacity
tortfeasor	one liable for a civil wrong, except regarding a contract or trust
uberrimae fidei	of the utmost good faith
ultra vires	beyond one's power
volenti non fit injuria	no wrong can be done to one who consents to the action

Index

Note: a numbered list of AIA standard forms appears at the front of the book.